Alexander Buchholz

Prospective Life Cycle Assessment of High-Temperature Superconductors for Future Grid Applications

Karlsruher Schriftenreihe zur Supraleitung Band 035

Eine Übersicht aller bisher in dieser Schriftenreihe erschienenen Bände finden Sie am Ende des Buches.

Prospective Life Cycle Assessment of High-Temperature Superconductors for Future Grid Applications

by
Alexander Buchholz

Scientific Publishing

Karlsruher Institut für Technologie
Institut für Technische Physik

Prospective Life Cycle Assessment of High-Temperature
Superconductors for Future Grid Applications

Zur Erlangung des akademischen Grades eines Doktor-Ingenieurs
von der KIT-Fakultät für Elektrotechnik und Informationstechnik des
Karlsruher Instituts für Technologie (KIT) genehmigte Dissertation

von Alexander Buchholz, M.Sc.

Tag der mündlichen Prüfung: 18. Januar 2022
Erster Gutachter: Prof. Dr.-Ing. Mathias Noe
Zweiter Gutachter: Prof. Dr. rer. nat. Armin Grunwald

Impressum

KIT Scientific Publishing

Karlsruher Institut für Technologie (KIT)
KIT Scientific Publishing
Straße am Forum 2
D-76131 Karlsruhe

KIT Scientific Publishing is a registered trademark
of Karlsruhe Institute of Technology.
Reprint using the book cover is not allowed.

www.ksp.kit.edu

Print on Demand 2022 – Gedruckt auf FSC-zertifiziertem Papier

ISSN 1869-1765
ISBN 978-3-7315-1194-6
DOI 10.5445/KSP/1000145972

Acknowledgement

This dissertation was written while I was working at the Institute for Technology Assessment and Systems Analysis (ITAS) at the Karlsruhe Institute of Technology (KIT) and could not have been done without help from several sides. At this point I would like to express my sincere gratitude to these people.

First of all, I would like to thank Prof. Dr.-Ing. Mathias Noe from the Institute of Technical Physics (ITEP), who agreed to supervise this work as the first reviewer. In the course of this work, the numerous discussions were a great help not only professionally but also methodically and were incredibly instructive.

Secondly, I would like to thank Prof. Dr. rer. net. Armin Grunwald from ITAS, who agreed to supervise this work as second reviewer. Another special thank you goes to my research group leader, Dr.-Ing. Marcel Weil, who provided me with advice and support during this work, but also showed confidence in my work. Without his expert guidance, this work could not have been carried out.

Furthermore, I would like to thank some industry partners who have helped me enormously with their expertise in the course of this work. In particular, I would like to thank Veit Große from THEVA and Albert Calleja from Oxolutia, who not only provided data on the production of high-temperature superconductors, but also answered many questions and, in the case of THEVA, even granted me access to the production facility. I would also like to thank Dr. Friedhelm Herzog of Messer, who provided important information and explanations on how the AmpaCity cooling system works. I would also like to thank Dr Klaus Ohla of Haynes International and Mr Lewis Shoemaker of Special Metals, both of whom provided information on the production of Hastelloy® C-276. I would also like to thank Mr Oliver Sauerbach from Westnetz, who explained the AmpaCity cable and the individual components to me in more detail during a personal inspection. In addition, I would like to thank all the members of the EU FastGrid project who gave me valuable feedback during the project.

Additionally, I would like to thank my colleagues at KIT. I would especially like to thank Dustin Kottonau and Eugen Shabagin from ITEP for their technical expertise, who provided valuable input on the superconducting cable and the cooling system.

Furthermore, I would like to thank all my colleagues at ITAS who have simplified and enriched my time as a doctoral student through both professional and informal discussions. I would especially like to mention Dr. Manuel Baumann, Merve Erakca, Paul Grünke and of course the RESET research group.

I would also like to say a special thank you to my parents Carola and Michael and my sister Tina. Without your constant support, this work would never have been possible. I would also like to thank my friends Céline, Lisa and Raphael who, especially during the pandemic, made sure that I didn't lose my head and could finish this work. Thank you for always being there for me.

And last but not least, I would like to thank my wonderful girlfriend Mandy, who has been an enormous mental help to me, especially in the final months of this work, and has always encouraged and supported me.

I would like to dedicate this work to my grandma Ursula as well as my grandpa Rudi, who unfortunately will not be able to see the end of this work. Thank you for everything!

Karlsruhe, October 2021

Alexander Buchholz

Abstract

High-temperature superconductors are a promising technology for energy transmission and distribution. Two important properties that describe the superconducting state are immeasurably low ohmic resistance and diamagnetism. Both properties allow superconductors to be used in a wide range of applications such as power cables, magnetic coils in fusion reactors, or fault current limiters.

However, superconductors must be cooled constantly to achieve their superconducting state. This requires a cooling system during the use phase. Therefore, it is important to analyse the environmental impacts of superconductors not only during the production phase but throughout their entire product life cycle.

In this study, the method of prospective life cycle assessment is used to analyse the environmental impacts of the production of high-temperature superconductors and their application in a superconducting medium voltage cable for energy distribution. Prospective life cycle assessment enables the quantification of environmental impacts throughout the entire life cycle of a product in its current state as well as in a potential future developed state. This study examines the contribution of the different processes along the life cycle of the product. For this purpose, the impact assessment method Environmental Footprint 3.0 is used which covers 16 different impact categories. Furthermore, the sum indicator Cumulative Energy Demand is used to analyse the total energy consumption throughout the entire process and supply chain. In addition to the contribution analysis, various scenario analyses are conducted to examine potential future developments in the superconductor production as well as in the designs of future superconducting cable systems in a prospective manner.

Two production techniques for high-temperature superconductors are analysed, inclined substrate deposition and inkjet printing. While the first is already commercially available, the second one is still in a research and development phase. Within the inclined substrate deposition, the silver layer and the gadolinium barium copper oxide layer have the highest share in the total environmental impacts with an average share of 39 % and 30 %, respectively. For the inkjet printing, yttrium barium copper oxide layer printing is the process with the highest environmental impacts with an average share of 48 %. In direct comparison, the average environmental impacts of the inkjet printing are five times as high as the ones of the inclined substrate deposition. Considering a potential future production development in terms of increased material and energy efficiency, the environmental impacts of the inclined substrate deposition can be further decreased. For the inkjet-printed tape, higher yield and more robust tape architecture are assumed to be future developments of the production process. These changes reduce the environmental impacts in all categories except for the resource use of minerals and metals. Due to the added silver stabilisation layer, the impacts in this category significantly increase.

A superconducting medium voltage cable is selected as a case study to assess the environmental impacts of a high-temperature superconductor application and to compare its impacts to conventional alternatives. The use phase is identified as the major driver of environmental impacts causing on average about 80% of the total lifetime impacts of a superconducting cable. Compared to a medium voltage conventional cable, the impacts of a superconducting cable are on average 34 % lower for a cable load of 0.7 granted that both systems use the same transformers. Thus, as long as the cable load is sufficiently high the superconducting cable provides ecological benefits compared to the conventional medium voltage cable. However, the superconducting cable does not provide environmental benefits when compared to a conventional high voltage cable regardless of the cable load.

In a scenario analysis, a changed transformer configuration for the superconducting cable system is analysed so that the transformation from 380 kV to 10 kV only requires one instead of two transformers. In terms of environmental impacts, such a system can perform better than a conventional high voltage cable system with two transformers leading to average savings of 25 % for a cable load of 0.7.

The use phase impacts of the superconducting cable systems are mostly caused by the liquid nitrogen consumption of the open cooling system. As the use phase is identified as the main source of environmental impacts, the usage of a closed cooling system is assessed in a prospective scenario analysis. Such a closed cooling system uses electricity to provide the required cooling capacity and can reduce the environmental impacts of a superconducting cable system by 10 %. If this closed cooling system is in addition operated with an entirely renewable energy mix, the environmental impacts can be decreased by 57 % compared to the original open cooling system.

This study shows that high-temperature superconductors can not only be an environmentally friendly alternative to conventional conductors but also identifies further potential to increase the environmental benefits during production and use phase. However, the environmental impacts of high-temperature superconductors strongly depend on various parameters such as the field of application, the cable load, or the design and configuration of the system components during the use phase.

Table of Contents

1 Introduction

During the 2015 United Nations Climate Change Conference, 196 parties approved a legally binding international treaty on climate change, the so-called Paris Agreement. The purpose of this treaty was to call for actions in order to mitigate climate change that is arguably the most important threat to the environment right now. Thus, these 196 parties agreed to limiting the global average temperature increase to well below 2°C while even attempting to keep it below 1.5°C [1].

According to the sixth assessment report of the Intergovernmental Panel on Climate Change, the estimated remaining carbon budget to achieve the 1.5°C goal with a likelihood of 83% is 300 Gt CO_2 [2]. In 2016, the global energy sector alone was responsible for 73.2% of the greenhouse gas emissions with a total amount of about 32-36 Gt CO_2 [3] [4]. In 2019, the German energy industry was responsible for 30.8% of the total greenhouse gas emissions with a total amount of 250 million tons of CO_2 [5]. Additionally, the energy-related emissions of other industry sectors caused up to 23.1% of the German greenhouse gas emissions in 2019. In the US, the electricity sector caused 25.0% of the 6.5 Gt CO_2 emissions in 2019 [6]. Additionally, about 500 million tons are caused by electricity consumption of the industry sector.

These numbers already show that the energy sector plays a crucial role in achieving the Paris Agreement goals. Therefore, actions must be taken in order to reduce the impact that the energy sector has on the global climate. In order to reduce the emissions of the energy sector there are a few strategies: Using renewable energy sources, electrification of fossil fueled sectors such as transport, and increasing the energy use efficiency. The increased use of renewable energy will result in a greater spatial discrepancy between energy production and energy consumption. Additionally, renewables are more volatile than fossil fuels, which requires a more flexible power grid. Furthermore, electrifying the transport sector would increase the energy consumption. According to the German Federal Ministry for the Environment, Nature Conservation and Nuclear Safety, the total energy consumption would increase by about 15 % if all combustion cars in Germany were replaced by electric cars [7]. For these reasons, the power grid will have to be expanded and reconstructed in the future.

However, this can lead to problems, especially in urban areas. Urban areas are responsible for around 75 % of the global primary energy supply consumption [8]. In addition, due to urban growth it can safely be assumed that this value will further increase in the future. However, there is only limited space in cities and thus the power grid must expand while also becoming more space efficient.

As a potential solution, superconducting cables can be used for future grid applications especially in urban areas. Superconducting cables have a much higher current carrying capacity and are much more compact than conventional cables [9]. Instead of conventional conductors with copper, these cables make use of high-temperature superconductors that provide unique properties.

Superconductors are materials that show two important properties: Immeasurably low DC electrical resistance and the magnetic field expulsion, the so-called Meissner Ochsenfeld effect. Superconductors were discovered in 1911 by Dutch physicist Heike Kamerlingh Onnes as they were analysing the resistance of pure mercury at low temperatures [10]. Kammerlingh Onnes discovered a sudden drop in resistance to immeasurably low values when cooling mercury below a temperature of 4.2K.

This discovery already sparked the idea of using superconductors to transmit electricity without ohmic losses. Superconductors can carry high currents up to the so-called critical current density J_c that depends on the critical temperature T_c and the critical magnetic field B_c. If one of these critical parameters is reached, the superconductor loses its superconducting property. To use superconductors for electricity transmission they need a constant cooling. However, the first superconductors had to be cooled using liquid helium, which is too expensive to be used in a commercial application.

In 1986 however, Bednorz and Müller discovered superconductivity in cuprates marking the discovery of the so-called high-temperature superconductors [11]. Their critical temperatures were significantly higher than the critical temperatures of any previously known superconductor [12]. A year later, the first high-temperature superconductor with a critical temperature in the range of the temperature of liquid nitrogen was discovered with yttrium barium copper oxide. As liquid nitrogen is much cheaper than liquid helium, this discovery enabled the use of high-temperature superconductors for energy transmission.

Due to the higher current carrying capacity and the low losses of a superconducting power cable, they can be used to make the energy grid more efficient and thus contribute to reducing the greenhouse gas emissions. However, as they need constant cooling, this cooling requires additional electricity consumption, which could potentially result in a rebound effect if the energy consumption of the cooling is higher than the loss savings. It is important to analyse the environmental impact of the use of superconducting cables in the power grid, not only in terms of greenhouse gas emissions but rather in a variety of environmental aspects. In addition to analysing various environmental aspects, it is also important to analyse the entire product life cycle and not only the production and use phase.

So far, the environmental impacts of the application of superconductors as well as of the superconductors themselves have not been analysed thoroughly. Often, studies qualitatively conclude that superconducting cables would have smaller environmental impacts by simply focusing on the reduced losses during the use phase without considering the other life cycle phases. Nishijima et al. analysed various energy applications of superconductors such as superconducting magnetic energy storage or direct current (DC) power lines [13]. However, while they state that superconducting cables would have a smaller environmental footprint than conventional cables, they did not provide any source or calculation for that claim. Baumann did analyse the usage superconducting magnetic energy storage [14]. They concluded that their environmental benefits stem from the resulting temporal decoupling of energy generation and consumption that has the capability of reducing fossil fuel emissions. While the statement in itself is correct, this effect is a property of energy storage technologies rather than of the superconductors themselves. Furthermore, they did not consider the required cooling during the use phase as well as the entire supply and production chain of the superconducting magnetic energy storage. Hawsey et al. concluded that superconducting cables have a lower environmental footprint than conventional cable due to their higher current carrying capacity [15]. However, they also did not consider the effect of the required cooling and the production of the cable and did not provide any calculations for the environmental benefits themselves.

While other studies provide life cycle assessments of superconducting applications and compare them to conventional alternatives, they focus only on greenhouse gas emissions. Hartikainen et al. conducted an LCA on superconducting magnetic energy storage, flywheels, and batteries as well as for superconducting and conventional cables [16]. They concluded that a superconducting magnetic energy storage using high-temperature superconductors has higher greenhouse gas emissions than a flywheel due to the lesser efficiency. However, in terms of the superconducting cable they concluded reductions in greenhouse gas emissions due to their higher efficiency at high market shares. In case of the cable, it is not clear whether they considered cooling during the use phase as well as the previous life cycle stages. Kamiya et al. also conducted an LCA on superconducting magnetic energy storage [17]. While they concluded that superconducting magnetic energy storage can reduce greenhouse gas emissions, they only considered energy flows and thus neglect any direct process emissions. In addition, they do not consider the required cooling of the superconductor. While focusing on only one environmental aspect can provide a first insight into potential environmental benefits of superconductors, potential environmental disadvantages might remain undetected.

However, there are also more detailed life cycle assessments for various superconducting applications. Lloberas-Vallas et al. analysed the environmental effects of a 15 MW superconducting synchronous generator in six environmental impact categories [18]. However, they only conduct a cradle-to-gate analysis the does not include the use phase of the generator. Nevertheless, in their study the environmental effect of the superconducting material is negligible in all considered categories except for eutrophication. Berti et al. analysed eleven environmental impacts of two superconducting 25 MVA transformer with different superconducting tapes in a cradle-to-grave approach [19]. They compared the superconducting transformers to a conventional conductor and conclude that the superconducting transformers have significantly lower impacts. However, while they did include the use phase of the transformers in their study, they only considered the losses of the transformers and not the required cooling and thus underestimate the impacts during the use phase. Marian et al. analysed an MgB_2 high voltage direct current superconducting cable [20]. However, in their study they only provide a contribution analysis of the different cable layers while also assuming that the cable is fully loaded throughout its entire lifetime. Thus, they analysed a rather unrealistic use phase scenario. Furthermore, they did not include the required cooling of the cable.

Therefore, to the knowledge of the author this study is the first to conduct a detailed LCA on the production of high-temperature superconductors. Two different production techniques on different technology readiness levels are examined. In addition to analysing the current superconductor tape production techniques, potential future developments of these techniques are analysed in a prospective LCA. In this prospective analysis, changes in material and energy efficiency of the production processes as well as potential future changes in tape architecture are considered. Furthermore, this study conducts an LCA on the application of these superconducting tapes in a 10 kV superconducting power cable. Besides covering multiple environmental impact categories, this study also covers the use phase of the cable by including various operational conditions as well as the required cooling of the cable. In addition, the results of the superconducting cable are compared to the environmental impacts of conventional cables. Furthermore, a prospective analysis of the superconducting cable is done by examining the potential future use of a different cooling system, changing transformer configurations, and analysing the effects of using an electricity mix entirely based on renewable energy sources. Thus, this study is the most detailed life cycle assessment on high-temperature superconductors and their future grid application up to date.

In the following, the method of life cycle assessment will be presented in chapter 2 by describing the four parts of a life cycle assessment. The life cycle assessment of a copper conductor wire will be used as an example. In chapter 3, two different techniques to produce high-temperature superconducting rare-earth barium copper oxide tapes will be analysed and compared. The first production technique is already on an industrial scale and commercially used, while the second one is an emerging technique that is still at laboratory scale. Within Chapter 4, a 1 km long, 10 kV, 2.31 kA superconducting cable will be compared to a conventional 10 kV and a conventional 110 kV cable. Furthermore, the effect of fluctuating loads, transformer configurations and cooling systems will be analysed. In chapter 5, a the work will be summarised and a conclusion as well as an outlook are provided.

2 The Method of Life Cycle Assessment

2.1 Introduction

Life cycle assessment (LCA) is a method that was developed in the US around 1970 [21] [22]. The basic idea of LCA is to quantify the environmental impact of a product or service not only during its production but over its entire life cycle in a systematic way [22] [23]. In addition to the design and production phase, this life cycle also includes the interlinked steps raw material extraction, packaging and distribution, use and maintenance phase, and disposal, recycling or reuse as it is shown in Figure 2.1.

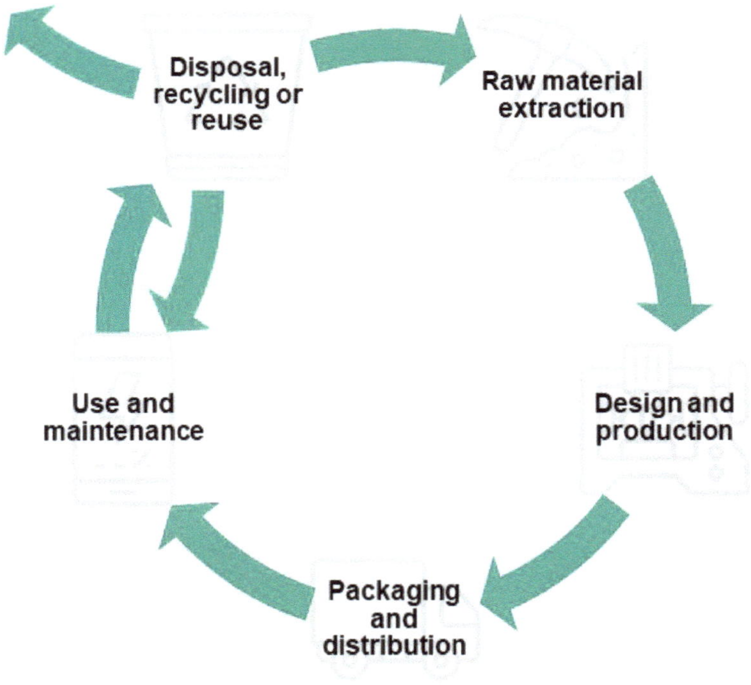

Figure 2.1: Typical scheme of a product life cycle (based on [24]).

In the early 1990s, the Society of Environmental Toxicology and Chemistry developed the first guidelines for conducting a life cycle assessment [25] [26]. Shortly after, the International Organization for Standardization dealt with the standardization of the individual methods developed [23]. Currently, there are two international standards for conducting an LCA:

- ISO 14040:2006: Environmental management – Life cycle assessment – Principles and framework [27]
- ISO 14044:2006: Environmental management – Life cycle assessment – Requirements and guidelines [28]

Figure 2.2 shows the interconnections of the four steps that are mandatory for each LCA according to these ISO standards as well as possible applications of an LCA:

1. Goal and scope definition
2. Life cycle inventory analysis
3. Life cycle impact assessment
4. Interpretation

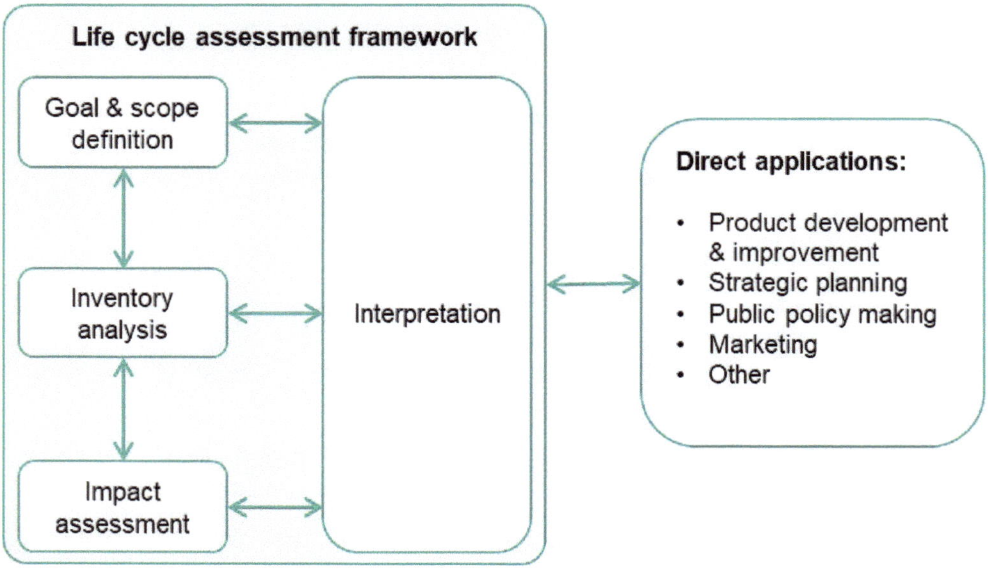

Figure 2.2: LCA framework according to the ISO 14040:2006 standard with the four mandatory steps as well as examples for direct LCA applications (based on [27]).

In addition to the ISO standards, the European Commission provided various International Reference Life Cycle Data System handbooks that serve as detailed guides for completing each of these steps in the European Context [29] [30] [31] [32] [33]. It is worth mentioning that an LCA is not a linear process, but iterative. New findings during the life cycle inventory analysis step can, for example, lead to adjustments of goal and scope afterwards. In the following, the mandatory steps are explained using the example of the production of a conventional copper conductor wire.

2.2 Goal and Scope Definition

2.2.1 Goal Definition

At the beginning of each LCA, the goal of the study is defined and thus the context of the study is provided. Here, the reasons for the study as well as its planned application are explained [27].

The reasons for the study describe the motivation for carrying out the LCA [30]. This then results in the requirements for data quality. This is because depending on the goals, data requirements can be higher in terms of completeness, reliability, and geographical, temporal and technological correlation while in other cases a lesser data quality is still sufficient to reach the goal of the study. Furthermore, the decision context is explained here. This affects the modeling of the inventory data in the following step, since depending on the context an attributional or a consequential LCA is performed. The former focuses on quantifying the environmental impact of an existing product system. The latter aims at assessing the potential impact changes due to a new product that is substituting another product on the market.

The planned applications, for example, can be a comparison of two products or an ecological hot spot analysis of a product. Furthermore, the chosen methods and impacts, potential assumptions, and the resulting limitations are described [30]. Impact limitations, for example, can occur if a one-dimensional impact such as the carbon footprint is chosen, since this only considers one facet of environmental impacts. Methodological limitations may include the use of site-unspecific data for the analysis of site-specific systems. Assumptions about the system under study may also limit the transferability of the results. These include assumptions about unusual conditions of use, such as the comparison of cable systems under the assumption of continuous full load.

Furthermore, during the goal definition the target audience is identified [27]. Thereby, it is also described how the results must be reported. This step is important to ensure that the study is sufficiently detailed for the objective and is internally consistent.

Example

For the example study of the copper conductor wire production, the goal must be defined first. In this chosen case, the goal is to analyse the carbon footprint of the production of the wire to identify the key processes that contribute the most to the carbon footprint. The example study is commissioned by the conductor manufacturer themselves. Thus, as they are the target audience the results will not be disclosed to the public, but only reported to the manufacturer. The aim is to provide recommendations on how to ecologically improve the conductor manufacturing regarding its greenhouse gas emissions. Hence, site-specific data for the life cycle inventory is required. Since only the carbon footprint is examined in this example study, this limits the results of the study to only one environmental impact. Thus, potential improvement with respect to the carbon footprint may cause harm in other environmental aspects that were not considered.

2.2.2 Scope Definition

The scope describes the examined object of the LCA in detail by identifying and explaining the following items:

- Examined product systems
- Functional units
- System boundaries
- Chosen allocation procedures
- Selected impact assessment methods and impact categories
- Initial data requirements

During the scope definition, the product system describes the processes that need to be analysed to achieve the goal of the study in more detail [30]. The product system can be divided into two parts - the foreground and the background system [34] [35]. The foreground system describes the part of the system where primary, site-specific data is required to be collected by the LCA practitioner. The background system describes the part of the system where averaged data can be used. Generally, it consists of secondary data, most commonly from life cycle inventory databases or published literature.

Additionally, the function of the product system must be identified [27]. The function of a product system is necessary to describe the analysed object of the LCA quantitatively and qualitatively. In general, this is done by using a functional unit that quantifies the function by asking questions like 'how much', 'how well', or 'for how long' to describe the function [30]. Answering these questions determines the so-called reference flow of the product system. The reference flow is the quantitative desired output of a product system. All other input and output flows are scaled in a way that the product system produces exactly one unit of the reference flow. Additionally, the reference flow is particularly relevant for the comparison of different product systems to ensure that the comparison is made on a common basis. Thus, the functional unit is a quantified description of the function of the product system, while the reference flow is the quantified amount of a manufactured product that is necessary to provide the function as described by the functional unit [36].

Furthermore, the system boundaries are defined in this step. Each LCA model describes a section of the real world that consists of man-made objects and processes (technosphere) and the natural environment (biosphere) [30]. Thus, the system boundaries are important for two reasons [35]. Firstly, they identify all the processes that are relevant to provide the functional unit. Thus, the system boundaries can be used to differentiate the product system from the rest of the technosphere. Secondly, the system boundaries examine where the interaction and exchanges between technosphere and biosphere takes place [30]. System boundaries can also include different aspects of the product life cycle. Depending on completeness, a distinction is made between "cradle-to-gate", if the product life cycle is only considered up to production, or "cradle-to-grave", if the use phase as well as the end-of-life are also considered [37].

Additionally, it is possible that the product system is a multi-functional system that results in more than one reference flow. In this case, all the input and output flows must be quantitatively distributed to these reference flows. This process of distribution is called allocation and can be done in several ways. For example, it can be performed physically based on mass or economically based on the financial value of the co-products. Each allocation method has its own advantages and disadvantages. The mass allocation is more logical in terms of material and energy flows. The economic allocation rather reflects the societal cause of the emissions as a higher monetary value of a product reflects a higher societal demand for this specific product [38]. The choice of allocation method thus has a decisive influence on the result of the LCA, as has already been shown in various studies [39] [40] [41]. Hence, it might be necessary to analyse the effect of the allocation method selection in a sensitivity analysis.

In addition, the scope describes the selected impact assessment method and impact categories. It is also determined if the LCA is done using a midpoint approach, am endpoint approach, or even a single-score indicator. All three of these approaches assess environmental impacts at specific points in a cause-effect chain. This cause-effect chain consists of an emission (such as CO_2), the resulting environmental mechanism (such as global warming potential), and the damage caused by this environmental mechanism (such as the extinction of species).

In a midpoint approach, the environmental impact of emissions is assessed in terms of environmental mechanisms within the cause-effect chain. [42]. At midpoint level, there are sets of indicators such as climate change or freshwater ecotoxicity [43]. Each indicator consists of a set of characterisation factors that reflect the relative importance of an emission or extraction compared to a reference emission or extraction [42]. Carbon dioxide (CO_2), for example, is the reference emission of the indicator climate change. The characterisation factors of this indicator now quantify the relative impact of each relevant emission in the category compared to CO_2. In general, the midpoint approach is based on scientifically robust methods and provides a detailed overview [43] [44]. However, since there is usually a variety of indicators, the midpoint approach might not always be able to answer a question like “Is product A better than product B?” in an easy manner [44]. The reason for this is that different indicators might favor a different product, which can make it harder to interpret the LCA results.

An endpoint approach assesses resulting damages caused by the environmental mechanisms. It converts the indicators of the midpoint impact categories into damage categories in areas of protection such as human health damage or ecosystem damage [43]. Similar to the midpoint approach, the endpoint approach uses damage factors that quantify the relative importance of each impact category [43]. Endpoint results are generally easier to comprehend. However, due to a more complex impact pathway modelling, an additional uncertainty is introduced to the results [42] [44]. Endpoint results can also be aggregated into single-score indicators to provide an even easier comprehensible result.

Single-score indicators can be applied at both midpoint and endpoint levels and aggregate different impact categories into a single value which enables easy comparisons between multiple product systems. However, a single-score indicator requires normalisation and weighting to aggregate results, which is a subjective process.

There are various impact assessment methods available for LCA. While some of them only cover midpoint indicators, others also provide endpoint indicators. Some of the most frequently used methods include ReCiPe [45] [46], the International Reference Life Cycle Data System of the Joint Research Center of the European Commission [30], its successor Environmental Footprint [47], or the Cumulative Energy Demand [48].

Furthermore, data quality requirements must be defined within the scope definition. Data quality covers three different aspects [30]:

- Accuracy
- Precision
- Completeness

Accuracy describes how well the data represents the product system as well as if the methodological approach is appropriate regarding the goal of the study. Data precision quantifies the uncertainty of the data. In general, data quality can be assessed by using a Pedigree matrix as well as a probability distribution function [49]. The Pedigree matrix identifies data improvement potential in terms of temporal, geographical, and technological quality, completeness, and reliability. A probability distribution function quantifies a potential spread of all input and output flows of a life cycle inventory.

Example

For the copper example, the cradle-to-gate approach is used to calculate the carbon footprint. This is because the goal of this example study is to provide recommendations to the manufacturer on how to make their production more environmentally friendly. Therefore, the use phase as well as the end-of-life of the copper wire production are not considered.

The function of the product system is to provide a specific length of an insulated copper wire with a cross section area of 2 mm^2. Hence, the functional unit of this product system is one metre of copper wire.

Figure 2.3 shows the system boundary of the copper wire production and locates the product system within the techno- and biosphere. The foreground system consists of the three processes that are relevant to provide the functional unit. These are the required copper mass, the copper wire drawing and the final insulation of the wire with polyvinylchloride. In this study, ecoinvent 3.5 is used as background database [50]. The background system provides all the necessary material and energy flows from the technosphere as well as potential elementary resource flows from the biosphere. As the end-of-life of the copper wire is out of the scope of this example study, waste treatment processes are outside of the system boundaries.

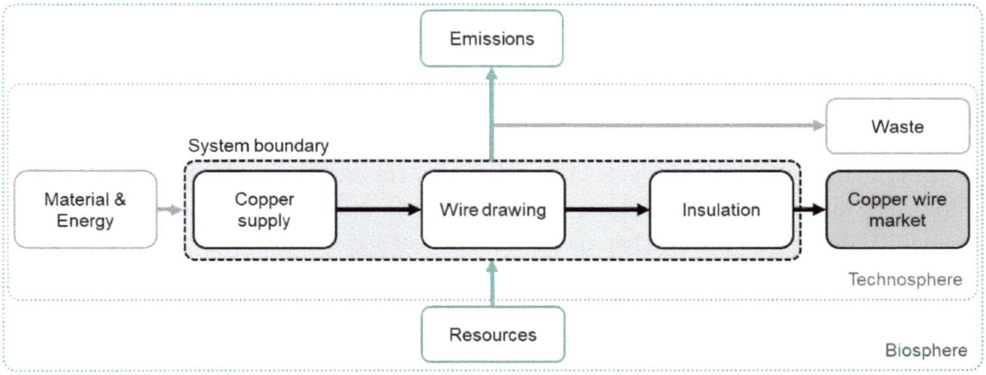

Figure 2.3: Exemplary representation of a product system for copper conductor production with the individual process steps and the system boundaries.

To quantify the carbon footprint of the copper conductor production, the impact assessment method Environmental Footprint 3.0 is selected. Environmental Footprint 3.0 includes the impact category climate change. The category indicator is the global warming potential of substances over the course of 100 years. The corresponding characterisation factors are based in the report of the Intergovernmental Panel on Climate Change [51]. Additionally, Environmental Footprint also includes substances with feedback mechanisms, such as methane [47] [52]. This allows a more accurate quantification of the carbon footprint. While Environmental Footprint includes various other impact categories, they are not considered within the scope of this example study.

Focusing only on one impact category, limits the significance of the study. Recommendations regarding a potential to decrease greenhouse gas emissions are possible. However, the same recommendations might lead to rebound effects in other impact categories that are not considered in this study.

According to the United Nations Environment Programme, the recycled content in copper products is 25-50 % [53]. Therefore, it is assumed that the copper used in this product system also contains secondary copper. The global copper market process in ecoinvent contains about 29 % of recycled copper and is selected as a provider process in this example study.

Because this study only serves as an example, site-unspecific average datasets are used in this model to calculate the carbon footprint. While there is no need for higher data quality, the results will still be reviewed and compared to other studies to check the validity of the results (see example in chapter 2.4.1).

2.3 Life Cycle Inventory Analysis

2.3.1 Data Collection

The collection of data for a life cycle inventory is an iterative process. In this process, information on all material and energy flows is collected for the individual processes of the product system [27]. The data is collected based on the initial goal and scope definition and data quality requirements for each unit process are identified.

In general, there are two types of data for a life cycle inventory [32]. Primary data refers to site-specific manufacturer information or measurements. Usually, the entire foreground system requires primary data [32]. Some parts of the examined product system are not required to have primary data. In these cases, the use of secondary data is appropriate. Secondary data refers to averaged or generic data sets from third parties, such as inventory databases or published literature values [30].

The required data type depends on the necessary data quality that was identified during goal and scope definition. However, data collection can lead to a better understanding of the examined product system. This, in turn, can lead to the identification of new limitations or data requirements that ultimately can result in an adjustment of the goal and scope definition.

Additionally, during the subsequent life cycle impact assessment step, unit processes with a low initial data quality might turn out to be an important driver of environmental impacts. In these cases, their life cycle inventories must be improved in terms of temporal, geographical, or technical quality or completeness.

Example

For the copper conductor wire production, averaged data sets for each of the identified unit processes in the product system are appropriate. Therefore, all processes are taken from the ecoinvent 3.5 database.

2.3.2 Data Calculation

The collected data of an LCA are presented in the form of so-called unit processes. Unit processes scale all input and output flows with respect to a reference flow. Hereby, the reference flow of a unit process is the primary output of the process. The remaining input and output flows are scaled to produce exactly one unit of the desired output.

Figure 2.4 shows the typical components of a unit process that can be clustered into four categories. There are inputs and outputs from and to the technosphere and the biosphere.

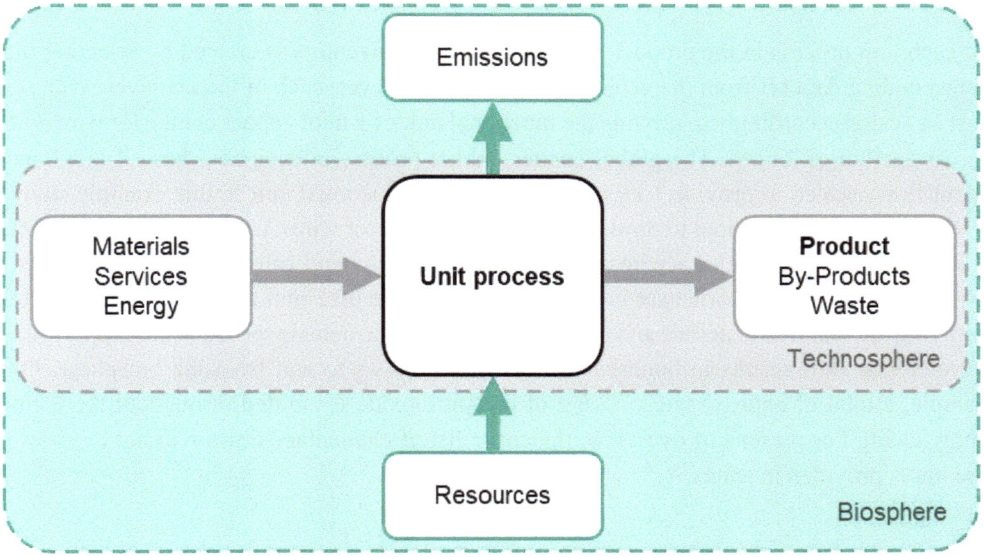

Figure 2.4: Schematic of a unit process with the two spheres, technosphere and biosphere, and their interactions (based on [54]).

Technosphere flows represent inter-industrial interaction within the process chain. Thus, technosphere inputs include all necessary precursor materials, energy flows as well as services like transportation. Technosphere outputs refer to all the flows that enter the technosphere coming from the unit process. These include the desired product or service (i. e. the reference flow), potential by-products as well as waste products.

Biosphere flows (also called elementary flows) refer to the interactions between the unit process and the natural environment. Biosphere inputs include all natural resources that are consumed by the unit process. Biosphere outputs refer to all substances emitted to air, soil and water by the unit process.

In some cases, a unit process might be multi-functional, meaning that in addition to the reference product there are also by-products. All input and output flows now have to be allocated to these various products. The allocation is done according to the procedure defined in the goal and scope of the study.

If a study is a pure life cycle inventory study, it ends with this step. In this case, only the life cycle inventory results are available. These describe all emissions and extractions caused by the reference flow. However, they do not address the resulting impact.

Example

For each unit process in the product system, a life cycle inventory is created by selecting the corresponding data set from the ecoinvent database. However, each of the ecoinvent data set must be scaled accordingly to provide the functional unit of 1 m of copper conductor wire with a cross section of 2 mm^2. The global copper market process in ecoinvent has all input and output flows scaled to provide 1 kg of copper. For the functional unit in this example study, 17.9 g of copper are required to produce the copper conductor wire. Additionally, these 17.9 g of copper must be drawn to a wire which consumes energy and causes emissions. At least, about 2 g of polyvinylchloride are necessary to provide the 0.25 mm thick insulation. Just by supplying the required materials and by consuming energy to actually produce the copper wire, the reference flow results in hundreds of elementary flows to and from the biosphere. For example, a total of approximately 77.6 g of carbon dioxide is emitted throughout the entire supply chain. For reasons of overview, the entire list of elementary flows will not be shown here but is provided in annex A.

2.4 Life Cycle Impact Assessment

2.4.1 Elements of Life Cycle Impact Assessment

The third step of an LCA, the impact assessment, is the evaluation of the potential environmental impacts based on the life cycle inventory results. Figure 2.5 shows the three mandatory steps of the life cycle impact assessment.

At first, the impact categories, the category indicators as well as the characterisation models are selected. In general, this first part is already included in the goal and scope definition of the study [36]. The impact category refers to positive or negative changes on the environment that are caused by anthropogenic emissions or extractions [55], such as climate change, freshwater ecotoxicity or ionising radiation. The category indicator is a quantifiable representation of the corresponding impact category at midpoint or endpoint level [36] [55] [56]. For the category climate change, the midpoint indicator is usually the infrared radiative forcing (W/m^2)

of a substance as this enables a representation of changes in the energy balance in the atmosphere. [46] [47]. The characterisation model represents an environmental cause-effect chain by describing the relation between life cycle inventory results and the category indicators [55]. The selection of impact categories, category indicators, and characterisation models answers the question “Which impacts are assessed?”

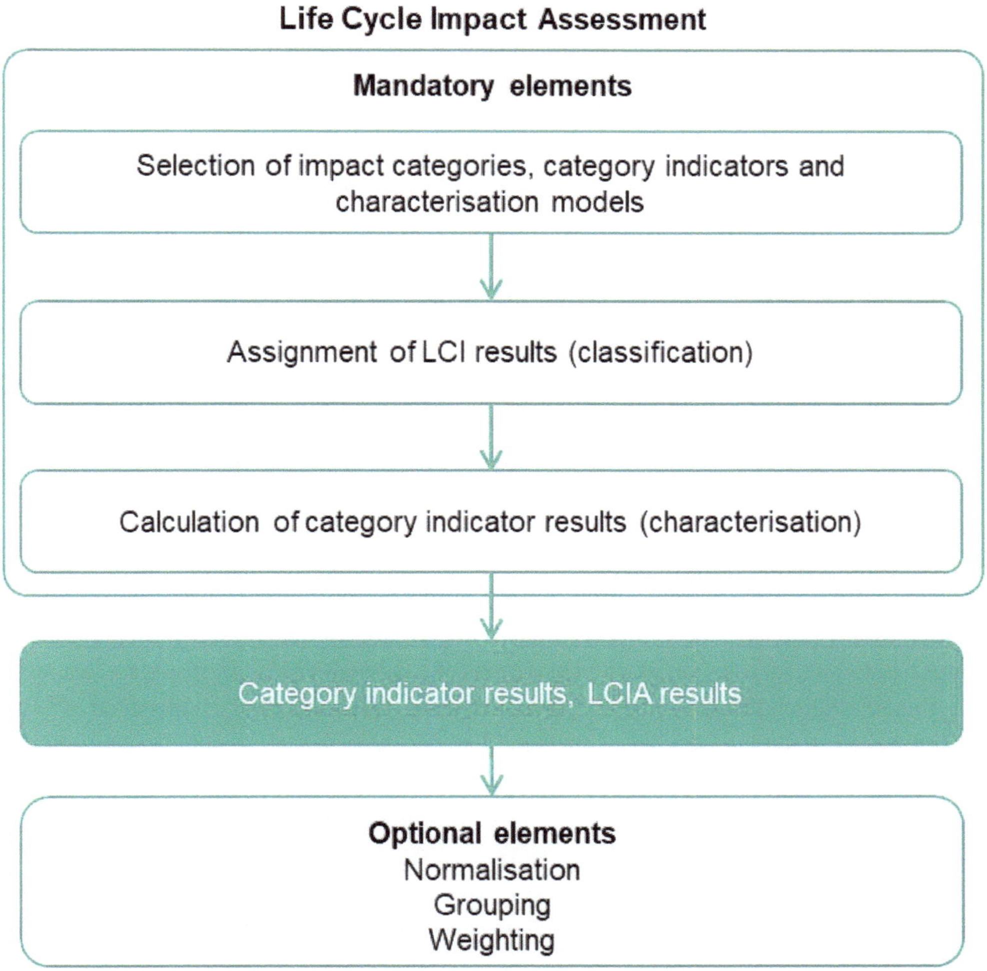

Figure 2.5: Elements of the life cycle impact assessment phase (based on [27]).

During classification, the elementary flows identified during the life cycle inventory step are assigned to one or more impact categories. For example, within the life cycle impact assessment method Environmental Footprint, carbon dioxide is assigned only to the impact category climate change. However, nitrogen dioxide affects different aspects of nature and is therefore assigned to several impact categories at once, including freshwater and terrestrial acidification and photochemical ozone formation. Thus, classification gives an answer to the question "Which impacts do the life cycle inventory results contribute to?" [55].

Subsequently, indicator values are calculated during characterisation. For this purpose, the life cycle inventory results converted to a common unit for each impact category. The conversion is done via characterisation factors which are derived from the selected characterisation model. The question answered by characterisation is "How much does each life cycle inventory result contribute to the impact?" [55].

There are also optional steps that can be taken. Normalisation aims at putting the life cycle impact assessment results into perspective by expressing them relative to a reference system such as per-capita averages. Weighting quantifies the importance of each impact category compared to each other. Grouping can be done to aggregate various categories. However, each of these optional steps can influence the results by subjective normalisation reference or weighting choices.

Figure 2.6 shows the mandatory steps of life cycle impact assessment along the entire environmental cause-effect chain for the impact category climate change. The life cycle inventory results describe all emissions and extractions of the product system. However, only the greenhouse gas emissions are assigned to the category climate change during classification. Each of these substances cause an infrared radiative forcing that changes the atmospheric heat adsorption. Thus, the infrared radiative forcing is the category indicator. The Intergovernmental Panel on Climate Change quantified the effect of each greenhouse gas in terms of its global warming potential over a time span of 100 years relative to carbon dioxide. These values are used as characterisation factors and are given in kilograms of CO_2 equivalents. Methane, for example, has a characterisation factor of 36.8 kg CO_2 equivalents. This value can be understood as "The infrared radiative forcing of 1 kg of methane has the same global warming potential over 100 years as the radiative forcing of 36.8 kg of carbon dioxide." During characterisation, these characterisation factors are assigned to the life cycle inventory results and since all substances share a common unit, the total indicator result can be calculated by simple addition. Based on damage factors, the environmental relevance of each indicator can ultimately be converted to the category endpoint to quantify potential damages to human health or ecosystems.

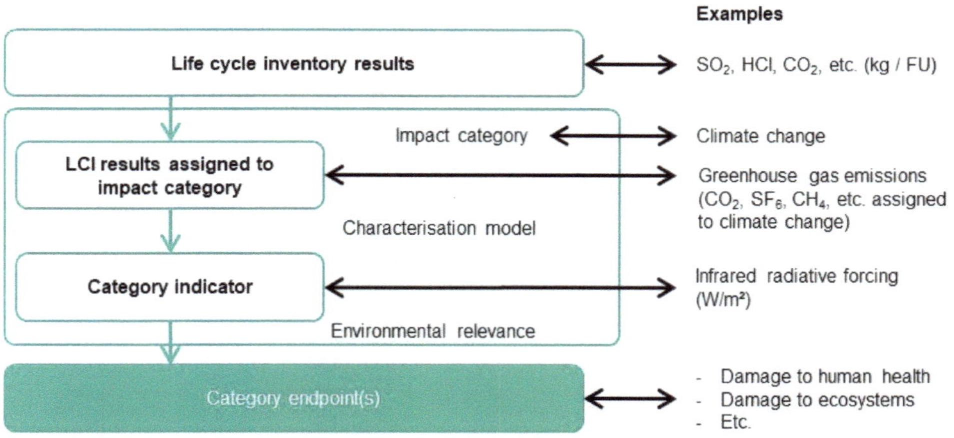

Figure 2.6: Concept of impact category indicators using the example of the impact category climate change (based on [28]).

Example

Table 2.1 lists all 28 substances that are emitted by the production of one metre of a copper conductor wire with a cross section of 2 mm² and that are assigned to the impact category climate change. For each substance, the life cycle inventory results as well as corresponding characerisation factors are given. By simple multiplication, the category indicator of each substance is calculated. Due to the common unit, the total category indicator can be calculated. In total, the production of one metre of copper conductor wire causes 87.44 g CO_2 equivalents.

The most influential unit process of the example study is identified using a contribution analysis. The copper supply is responsible for about 85 % of the entire carbon footprint. Of these 85 %, the majority (about 80 %) are caused by primary copper production, while the rest is caused by secondary copper. That means, that secondary copper, while accounting for roughly 28 % of the entire copper mass, causes a much smaller carbon footprint.

A comparison with a study by the German Copper Institute shows that the result of this sample calculation is within a plausible range [57]. In the study of the Copper Institute, a non-insulated copper wire with a cross-sectional area of 1 mm² was analysed, whereby a result of 27.6 g CO_2 equivalents was calculated for the climate change category. Extrapolated to a cross-sectional area of 2 mm², this would correspond to 55.2 g CO_2 equivalents per metre. In addition to the fact that insulation was not taken into account in the study by the Copper Institute, the differences may also be due to the different life cycle impact assessment methods used.

Table 2.1: List of all substances emitted by the copper conductor wire product system that are relevant to the category climate change including the life cycle inventory results, the characterisation factors as well as the individual and the total category indicator. The copper conductor wire has a copper mass of 17.9 g.

Substance	LCI result	Characterisation factor	Category indicator per functional unit
Bromethane, Halon 101	1.E-15 kg	3.0 kg CO_2eq./kg	3.0E-15 kg CO_2eq.
Bromochlorodifluoromethane, Halon 1211	1.8E-10 kg	2070.0 kg CO_2eq./kg	3.7E-07 kg CO_2eq.
Bromotrifluoromethane, Halon 1301	2.1E-10 kg	7150.0 kg CO_2eq./kg	1.5E-06 kg CO_2eq.
Carbon dioxide, fossil	0.07 kg	1.0 kg CO_2eq./kg	7.3E-02 kg CO_2eq.
Carbon dioxide, from soil or biomass stock	1.0E-04 kg	1.0 kg CO_2eq./kg	1.0E-04 kg CO_2eq.
Carbon dioxide, to soil or biomass stock	3.4E-08 kg	-1.0 kg CO_2eq./kg	-3.4E-08 kg CO_2eq.
Carbon monoxide, fossil	2.6E-04 kg	1.6 kg CO_2eq./kg	4.0E-04 kg CO_2eq.
Carbon monoxide, from soil or biomass stock	4.6E-08 kg	1.6 kg CO_2eq./kg	7.2E-08 kg CO_2eq.
Chlorodifluoromethane, HCFC-22	1.8E-09 kg	2110.0 kg CO_2eq./kg	3.9E-06 kg CO_2eq.
Chloroform	2.9E-10 kg	20.0 kg CO_2eq./kg	5.7E-09 kg CO_2eq.
Dichloromethane, HCC-30	3.7E-09 kg	11.0 kg CO_2eq./kg	4.1E-08 kg CO_2eq.
Dichlorodifluoromethane, CFC-12	3.3E-12 kg	11500.0 kg CO_2eq./kg	3.8E-08 kg CO_2eq.
Dichlorofluoromethane, HCFC-21	3.1E-14 kg	179.0 kg CO_2eq./kg	5.6E-12 kg CO_2eq.
Dinitrogen monoxide	1.6E-05 kg	298.0 kg CO_2eq./kg	4.8E-03 kg CO_2eq.
Methane	3.9E-10 kg	36.8 kg CO_2eq./kg	1.4E-08 kg CO_2eq.
Methane, fossil	2.0E-04 kg	36.8 kg CO_2eq./kg	7.3E-03 kg CO_2eq.
Methane, from soil or biomass stock	3.2E-09 kg	36.8 kg CO_2eq./kg	1.2E-07 kg CO_2eq.
Methane, non-fossil	4.3E-05 kg	34.0 kg CO_2eq./kg	1.5E-03 kg CO_2eq.
Methyl acetate	7.7E-15 kg	3.0 kg CO_2eq./kg	2.3E-14 kg CO_2eq.
Methyl formate	2.7E-13 kg	712.0 kg CO_2eq./kg	1.9E-10 kg CO_2eq.
Monochloromethane, R-40	1.2E-09 kg	15.0 kg CO_2eq./kg	1.8E-08 kg CO_2eq.
Nitrogen fluoride	3.1E-18 kg	17900.0 kg CO_2eq./kg	5.5E-14 kg CO_2eq.
Tetrachloromethane, R-10	5.0E-10 kg	2020.0 kg CO_2eq./kg	1.0E-06 kg CO_2eq.
Tetrafluoromethane, R-14	1.2E-08 kg	7350.0 kg CO_2eq./kg	8.7E-05 kg CO_2eq.
Trichlorofluoromethane, CFC-11	3.4E-14 kg	5350.0 kg CO_2eq./kg	1.8E-10 kg CO_2eq.
Trifluoromethane, HFC-23	9.9E-12 kg	13900.0 kg CO_2eq./kg	1.4E-07 kg CO_2eq.
Sulfur hexafluoride	7.4E-09 kg	26100.0 kg CO_2eq./kg	1.9E-04 kg CO_2eq.
Volatile organic compounds (VOC)	6.5E-08 kg	4.2 kg CO_2eq./kg	2.7E-07 kg CO_2eq.
		Total =	**87.44 g CO_2eq.**

2.4.2 Impact Assessment Methods

As noted previously, there is a wide range of impact assessment methods. These impact methods differ in the set of impact categories they contain as well as the underlying characterisation. Two life cycle impact assessment methods are presented in more detail below: Environmental Footprint 3.0 [47] and the sum indicator Cumulative Energy Demand [48].

2.4.2.1 Environmental Footprint

The Environmental Footprint method is derived from the International Reference Life Cycle Data System method, which was developed by the European Commission in2007 and published in 2010. Thus, Environmental Footprint uses the same nomenclature as International Reference Life Cycle Data System [47]. The first version of the Environmental Footprint was published in 2013 and has been continuously developed since then [58]. In 2018, Environmental Footprint 2.0 and later that same year Environmental Footprint 3.0 were released [47]. Environmental Footprint 3.0 includes 16 different impact categories. Additionally, Environmental Footprint provides a recommendation level for each category based on the scientific basis. These recommendation levels are:

- Level I: recommended and satisfactory
- Level II: recommended but in need of some improvements

- Level III: recommended, but to be applied with caution

All 16 impact categories are described in more detail below [47]. Furthermore, their respective recommendation level is given.

Acidification, terrestrial and freshwater (Level II)

The impact category of acidification was taken directly from the previous International Reference Life Cycle Data System method and not changed further. It quantifies the weighted sum of all accumulated exceedances of a critical load in an area of interest [59] [60]. Acidification is mainly caused by emissions of NH_3, NO_2 and SO_x. The reference unit of the characterisation factors is moles of hydrogen equivalents (mol H+ equivalents) per unit of mass.

Cancer human health effects and non-cancer human health effects (Level III)

These two impact categories are based on the USETox® Model 2.1 and quantify the emission of toxic substances into nature [61]. The distinction is made on the basis of carcinogenic substances. The reference unit for human ecotoxicity is the human comparative toxic unit, which reflects an estimated increase in mortality (in cases per kg) of the entire human population per unit mass of a chemical substance emitted.

Climate change (Level I)

For the quantification of greenhouse gas emissions, the characterisation factors of the Intergovernmental Panel on Climate Change 2013 were adopted [51] and adapted by the Joint Research Center under certain circumstances [47]. The global warming potential was selected for a time horizon of 100 years and carbon feedbacks of individual substances were also considered.

Ecotoxicity freshwater (Level III)

Like human ecotoxicity, freshwater ecotoxicity is based on the USEtox® model [61]. The reference unit here is the comparative toxic unit for ecosystems, which reflects the estimated potentially affected fraction of species integrated over time and volume, and the mass of chemical emitted (potentially affected fraction of species * m^3 * day * kg^{-1}).

Eutrophication freshwater and marine (Level II)

Eutrophication describes the overloading of ecosystems with nutrients, resulting in the death of plants and the resulting drop in oxygen concentration. The two aquatic impact categories of eutrophication are based on the EUTREND model [62]. While phosphorus is the limiting factor in freshwater environments, nitrogen is the limiting factor in marine waters. For this reason, the respective reference units of the two categories are kg P eq. and kg N eq. respectively.

Eutrophication terrestrial (Level II)

Terrestrial eutrophication is based on the same model as acidification and thus on the accumulated exceedances of a critical load [59] [60]. However, in contrast to acidification, moles of nitrogen equivalents (mol N eq.) of the impact indicator is used as the reference unit.

Ionising radiation – human health (Level II)

This impact category was adopted in its entirety from International Reference Life Cycle Data System. It quantifies the release of radionucleides into the environment that are harmful to human health [63]. Since the unit of elementary flows of radionucleides in the International Reference Life Cycle Data System is given as kBq and Environmental Footprint had adopted the nomenclature of the International Reference Life Cycle Data System, this category uses kBq U-235 eq as the reference unit of the characterisation factors [47].

Land use (Level III)

Unlike its predecessor method International Reference Life Cycle Data System, the Environmental Footprint method uses the land use indicator value calculation model called LANCA as the basis for evaluating land use impacts [64]. The LANCA model provides five indicators of soil use: erosion resistance, mechanical filtration, physiochemical filtration, groundwater recharge, and biotic production. The Joint Research Center calculates a single score index by aggregating these five indicators [65]. This index also serves as the reference unit for this impact category, which is thus expressed in points.

Ozone depletion (Level I)

For the assessment of the ozone depletion potential, characterisation factors from the assessment report of the world meteorological organization were used and extended by characterisation factors of the ReCiPe 2008 method [66] [67]. The category quantifies emissions of substances that deplete the stratospheric ozone layer and are controlled according to the 1987 Montreal Protocol. A kilogram of CFC-11 eq. was chosen as the reference unit for the characterisation factors.

Photochemical ozone formation – human health (Level II)

Under the influence of sunlight, photochemical ozone is formed in the troposphere from non-methane volatile organic compounds, which is harmful to human health. In this impact category, the characterisation factors were adopted from the International Reference Life Cycle Data System method based on the model used in ReCiPe [68]. The reference unit here is kg non-methane volatile organic compounds equvalents.

Resource use mineral & metals and energy carriers (Level III)

Resource consumption is divided into two categories. Minerals and metals are assessed on the basis of the abiotic resource depletion potential ultimate reserves version [69]. The reference unit used is kg antimony equivalent (Sb eq.) per kg extracted. The assessment of fossil fuel extraction is based on the same model, but due to their simultaneous function as energy sources, they are treated as a separate category. Here, the unit of the characterisation factors is MJ.

Respiratory inorganics (Level I)

This impact category evaluates the change in mortality due to particulate matter emissions and is based on a task force model for particulate matter by the United Nations Environment Programme and the Society of Environmental Toxicology and Chemistry [70]. As a reference unit of the characterisation factors, the disease incidences per emitted kg is used here.

Water use (Level III)

Water consumption in the Environmental Footprint method is calculated based on the Available Water Remaining model called AWARE [71]. This is a scarcity-adjusted water use, quantifying the relative amount of water available per area once the water demand of humans and aquatic ecosystems is met. The reference unit of the characterisation factors is m^3 deprived per m^3.

2.4.2.2 Cumulative Energy Demand

Cumulative Energy Demand differs from other impact assessment methods as it does not quantify a direct environmental impact of a product system. It rather quantifies the primary energy consumption over the entire product life cycle of a product system. For this purpose, the "energy harvested" approach is used, in which the total amount of energy sources provided for human use is quantified [48]. Both, renewable and non-renewable energy sources are considered. While renewable energy sources include biomass, water, wind, solar, and geothermal, non-renewable energy sources include fossil fuels, nuclear energy, and non-renewable biomass such as primary forests. The reference unit of the characterisation factors is MJ in all cases. Cumulative Energy Demand is an easily comprehensible single point indicator. Thus, it is often used in simple LCAs for communication with stakeholders.

2.4.3 Uncertainty Analysis

In addition to the mandatory steps of a life cycle impact assessment, other measures can be performed to assess the results. Uncertainty analysis is one of these measures and aims to investigate how uncertainties in the data and assumptions affect the robustness of the results.

2.4.3.1 Monte Carlo Simulation

An example of such an uncertainty analysis tool is Monte Carlo simulation. In Monte Carlo simulations, repeated random samples are drawn from a distribution to calculate the results [72].

In LCA, all input and output flows of a product system can be described by a probability distribution function instead of a fixed value like the average or the median. Monte Carlo simulations iteratively draws a random value from each probability distribution to calculate the indicator results. This process is performed in several, mostly thousands of iterations [73]. Contrary to using the fixed values, this process results in a probability distribution function for the category indicator itself. The benefit of this method is that the influence of data-inherent uncertainties can be quantified. The resulting probability distribution reflects the scatter of the results due to these uncertainties. The smaller the scatter, the more robust the results.

The probability distribution functions in life cycle inventories often given in the form of a log-normal distribution, which is because this probability distribution function always produces positive values [74]. However, it is important to note that the choice of the probability distribution function can have a significant impact on the life cycle impact assessment results and there has been no systematic study in which probability distribution function is the most appropriate for life cycle inventory data [74] [75].

2.4.4 Sensitivity Analysis

Sensitivity analysis is another method to better understand and interpret life cycle impact assessment results. In contrast to the uncertainty analysis, however, not the effects of all input parameters on the result are examined, but rather what influence individual parameters have on the overall system [28]. This method can be used to examine the effects of assumptions made and as well as potential technology changes [76]. There are various methods for sensitivity analysis. Among others there are tornado diagrams, scenario analysis, one-way sensitivity analysis, or critical error factor [77]. The first two mentioned methods will be presented in more detail in the following.

2.4.4.1 Tornado Diagrams

In the tornado diagram, the change in output parameters due to similar changes in individual input parameters is depicted. Here, one input parameter at a time is changed by the same factor, such as a 10 % increase. Meanwhile all other input parameters are kept constant [77]. The relative change in output parameters due to the relative change in input parameters is then plotted as a bar graph. The name of this method comes from the fact that the most sensitive input parameter has the widest bar and is listed at the top. All the other input parameters are plotted in descending order. The resulting diagram resembles an inverted triangle or a tornado in its shape.

2.4.4.2 Scenario Analysis

In scenario analysis, potential future changes in input parameters are used as the basis for calculations. Here, for example, different technologies, changes in the supply chain, efficiency improvements and similar parameters can be calculated [76] [77]. In contrast to the tornado diagram, it is not so much the influence of individual parameters on the system that is examined, but rather the influence of the various assumptions regarding processes, system boundaries or allocations.

2.5 Interpretation

The final step of an LCA is to jointly interpret the life cycle inventory and life cycle impact assessment results to reach conclusions consistent with the goal and scope definition, explain any limitations, and make recommendations. Here, the results should be reflected in an understandable, complete, and coherent manner. In this context, it should be kept in mind that LCA results represent potential environmental impacts and thus do not allow for predictions of actual impacts or damages.

<u>Example</u>

The contribution analysis identified the required copper as a main contributor to the carbon footprint. However, most of the the impact can be allocated to primary copper that makes up for 72 % of the required copper mass. It is therefore recommended that the copper conductor manufacturer uses a higher amount of secondary copper to reduce their carbon footprint. Using 100 % secondary copper would reduce the greenhouse gas emissions from 87.4 g CO_2 eq. to 30.9 g CO_2 equivalents. Hence, a significant reduction of the carbon footprint can be achieved.

2.6 Prospective Life Cycle Assessment

Prospective LCAs often address emergent technologies in early stages of development regarding their potential future environmental performance [78]. In addition, the effect of decisions regarding future strategies, such as energy pathways, can be assessed [79] [80]. This results in special features for an LCA. The examined technologies are those that are often still at the laboratory scale or barely entered the market yet [81]. Additionally, the performance of technologies under changing future circumstances can be analysed. Examples for such changed future circumstances could be a future electricity mix entirely from renewable sources, different production techniques, or varied system components during the use phase.

These features offer the possibility to identify potentials for ecological improvement or more environmentally friendly alternatives already at an early stage of technology development. However, scaling effects are expected, which must be considered especially in comparison with already established technologies. Such scaling effects could for example include a higher yield during production or a better efficiency during the use phase when compared to the earlier development stages of the technology. These scaling effects introduce uncertainty into the analysis, as attempts are often made to model a future and more advanced version of the technology [78]. Thus, prospective LCA always requires scenarios about possible future developments [81]. Here, the know-how of the developers can be decisive to depict realistic scenarios.

Furthermore, in many cases the data basis for emergent technologies is scarce or non-existent. Life cycle inventory data sets often need to be compiled from scratch. A temporal mismatch of the foreground and background data can also occur under certain circumstances [81].

Thus, the three main challenges of a prospective LCA are comparability with already established technologies, uncertainty, and the data basis. For this reason, it is recommended that both uncertainty and sensitivity analysis be performed in any prospective LCA to make the results more robust and relevant.

2.7 Tools of Life Cycle Assessment

2.7.1 Software: openLCA

Several softwares have already been established for carrying out a life cycle assessment. For example, there are the commercial softwares SimaPro, GaBi, and Umberto [82] [83] [84]. In this dissertation, the software openLCA of the German company Greendelta is used [85]. This is a free open-source software, which is based on the idea of Andreas Ciroth, Jutta Hildenbrand, and Michael Srocka and was developed in 2006. Since then, the software is in constant development. Version 1.10.3 of openLCA is used in the context of this dissertation.

2.7.2 Background Database: ecoinvent

Background databases are essential for carrying out an LCA, as it is virtually impossible to collect primary data for all products in the upstream chain for a product system. These background databases provide aggregated life cycle inventory data for a wide variety of processes, allowing an LCA practitioner to focus on foreground data relevant to their study.

This dissertation uses the ecoinvent database, which contains international life cycle inventory datasets from a wide variety of industrial sectors, including among others agriculture, energy supply, transportation, or waste treatment [86]. With over 18,000 life cycle inventory datasets, ecoinvent currently represents the worldwide largest background database.

Ecoinvent provides its background databases with different allocation systems. All calculations performed in this dissertation are based on the cut-off system model of ecoinvent 3.5 [87].

This is based on the idea that the primary production of all materials is allocated to the primary user of those materials. This means that in the case of recycled materials, the primary producer does not receive credits for providing recyclable materials. As a result, when recycled materials are used, only the burden of the recycling process needs to be assessed. The material itself is burden-free. Furthermore, waste producers do not receive credits for recycling or re-use of products generated during waste treatment. As an example, heat from the incineration of waste could be used burden-free, while the environmental impacts of incineration are allocated to the waste producer.

However, it is also important to note that for multi-functional product systems that have common by-products, the allocation of input flows is done by the authors of the individual life cycle inventory data sets. If necessary, it must be checked for each life cycle inventory data set of a product system which allocation method is used.

3 Life Cycle Assessment of Rare Earth Barium Copper Oxide High-Temperature Superconductor Tape Production

3.1 High-Temperature Superconductor Tape Production Techniques

There are two major types of high-temperature superconductors (HTS) [88]: Firstly, there are the bismuth strontium calcium copper oxides. These bismuth-based conductors are produced using the so-called powder-in-tube method. In this process, filaments of the bismuth strontium calcium copper oxides material are embedded in a matrix, e.g. of silver. Secondly, there are rare earth barium copper oxides (REBCO) high-temperature superconductors which are seen as the preferred option for future power applications due to their higher critical current density and better mechanical properties [89] [90].

In rare earth barium copper oxide superconductors, a superconducting rare earth barium copper oxide layer is applied to a metallic substrate, which promotes the biaxial growth of the superconducting layer [91]. However, there are different production processes to apply the individual layers. Firstly, a suitable substrate base must be prepared to provide the necessary texture that ensures the correct orientation of the growth of the following layers. The four methods typically used for this are: Rolling-assisted biaxially-textured substrate (RABiTS™) [92], ion-beam-assisted deposition [93], inclined substrate deposition [90], and inkjet printing [94]. In addition, there are also different methods for applying the superconducting layer. A distinction is made between three techniques: physical vapor deposition [95], metal-organic chemical vapor deposition [96] and chemical solution deposition [97].

Figure 3.1 shows a schematic of the RABiTS™ production process and a typical high-temperature superconductor architecture. The RABiTS™ process uses biaxially textured nickel-tungsten as a substrate [91] [92]. The texturing is created by the deformation and the subsequent recrystallisation of nickel. The deformation is achieved by cold rolling of the substrate, while the recrystallisation is done in a reel-to-reel furnace [91]. A cerium oxide buffer layer is also applied to this substrate, which preserves the structure of the substrate. Additionally, this buffer layer prevents a reaction of the substrate with the subsequent layers [98]. As the different layers have different thermal expansions, cracks can occur during production. To mitigate this, an yttria-stabilised zirconia (ZrO_2) layer is applied via a physical vapour deposition like pulsed laser deposition or a chemical vapour deposition [92] [99] [100]. This is then coated with the actual superconducting yttrium barium copper oxide film.

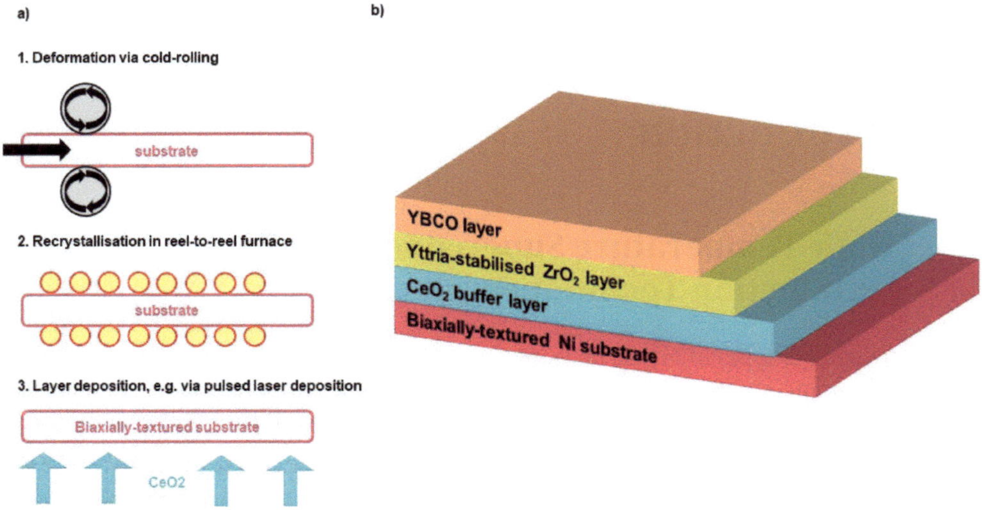

Figure 3.1: Schematic of a) the RABiTS™ production steps of deformation, recrystallisation and layer deposition and of b) the resulting architecture of a rare earth barium copper oxide high-temperature superconductor tape produced with RABiTS™ (based on [92] and [101]).

The ion-beam-assisted deposition method is a combination of a physical vapour deposition method, such as pulsed laser deposition, and ion bombardment [102]. This allows them to be directed and corroded if misoriented, thus controlling the growth orientation of the applied layer. The ion beam is usually a beam of argon ions [103]. In Figure 3.2, the process of ion-beam-assisted deposition and the typical resulting tape architecture are illustrated. In contrast to the RABiTS™ process, the biaxial texturing is not created by the substrate, but by the yttria-stabilised zirconia layer, which is vapour-deposited directly onto the substrate using ion-beam-assisted deposition. A cerium dioxide buffer layer and the superconducting rare earth barium copper oxide layer are then applied to this yttria-stabilised zirconia layer by pulsed laser deposition [93].

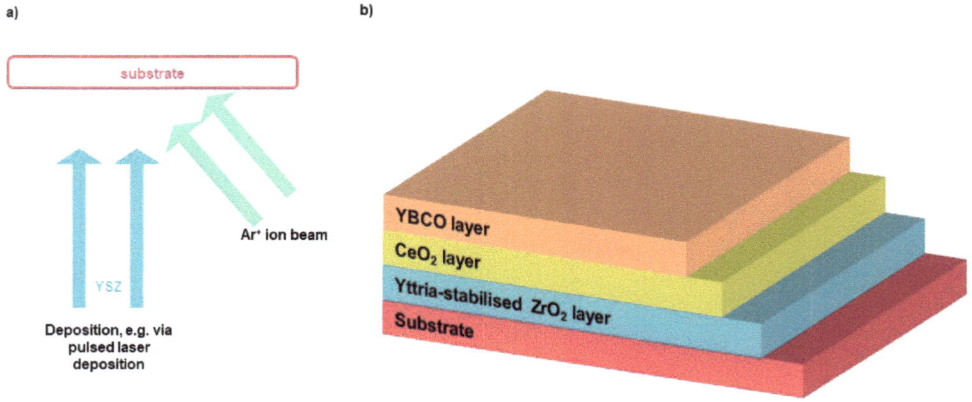

Figure 3.2: Schematic of a) the process of ion-beam-assisted deposition with its argon ion beam and of b) the resulting architecture of a rare earth barium copper oxide high-temperature superconductor tape produced with ion-beam-assisted deposition (based on [93] and [101]).

In the inclined substrate deposition method, a biaxially textured magnesium oxide layer is evaporated onto a metallic substrate. Figure 3.3 shows schematically the process of inclined substrate deposition as well as the architecture of the high-temperature superconductor tape as it is analysed in this study. The substrate is made of electropolished Hastelloy® C-276, a nickel-chromium-molybdenum-tungsten alloy, which is placed in a reel-to-reel system at a certain angle to the evaporated MgO to control the growth direction and the biaxial texture of the MgO buffer layer [90] [104]. The MgO layer applied via the inclined substrate deposition process serves as a crystalline base for the rare earth barium copper oxide layer and ensures that the rare earth barium copper oxide crystals grow in the desired orientation. The rare earth (RE) barium copper oxide layer is the actual current carrying layer. It is a RE-$Ba_2Cu_3O_{7-\gamma}$ ceramic, with yttrium (YBCO), gadolinium (GdBCO), or dysprosium (DyBCO) mostly used as the rare earth material. A metallisation is applied over the rare earth barium copper oxide layer, which consists of silver, for example. It serves as an electrical contact and as a chemical protection [105]. In some cases, a shunt is also applied, which is made of copper or Hastelloy® C-276, for example. The shunt increases the electrical and mechanical stability of the tape conductor.

Compared to other manufacturing processes, inclined substrate deposition has the property that the rare earth barium copper oxide layer thickness can be increased without significant decrease of the critical current density. Conductors with DyBCO with a current carrying capacity of more than 1000 A/cm and a layer thickness of nearly 6 µm have already been produced using this process [90]. Furthermore, the deposition rate of the inclined substrate deposition method is higher compared to ion-beam-assisted deposition which is beneficial regarding mass production [100] [106].

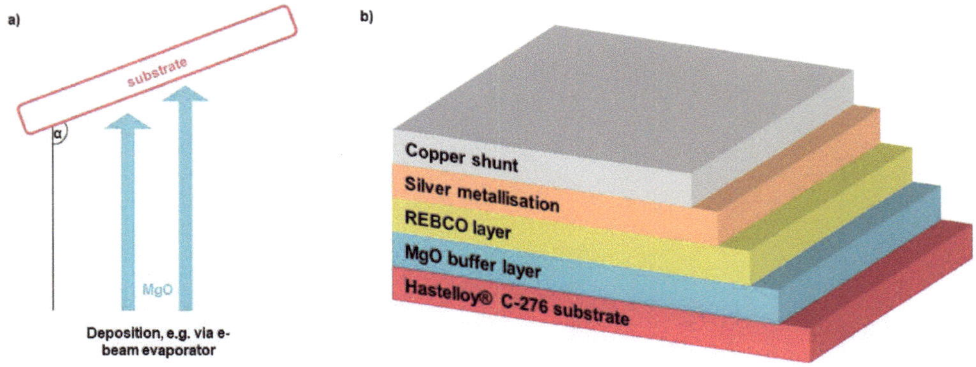

Figure 3.3: Schematic of a) the process of inclined substrate deposition and of b) the resulting architecture of a rare earth barium copper oxide high-temperature superconductor tape produced with inclined substrate deposition (based on [90] and [101]).

Another method for producing superconductors is inkjet printing. Compared to the other methods, inkjet printing has low costs and can also reduce alternating current (AC) losses due to the production of multifilamentary conductors [107]. In this process, a metallic textured substrate is used and the individual layers are applied in the form of a chemical solution deposition. Figure 3.4 schematically represents the production process as well as a resulting architecture as examined in this study.

An ink solution is prepared for each of the individual layers, which is then printed onto the substrate or the underlying layer in the form of individual drops using inkjet printing. The printed film is then dried by means of pyrolysis. A buffer layer consisting of a cerium-zircon ink is first printed on the sapphire substrate, which provides a more suitable lattice constant for the subsequent yttrium barium copper oxide layer to grow on [107]. The architecture investigated in this study also features a current flow diverter printed on the yttrium barium copper oxide layer to protect the tape in the event of a quench [108].

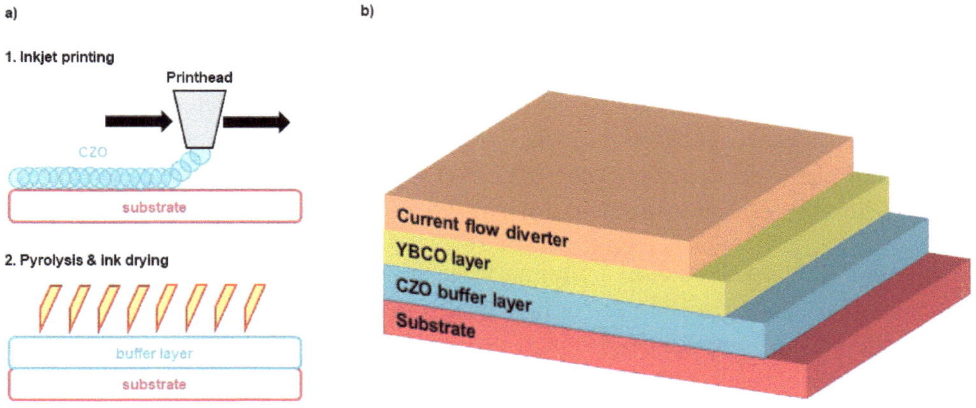

Figure 3.4: Schematic of a) the process of inkjet printing with the subsequent pyrolysis and of b) the resulting architecture of a rare earth barium copper oxide high-temperature superconductor tape produced with inkjet printing (based on [94] and [109]).

In the following, the architecture as well as the materials of the high-temperature superconductor tapes produced by the inclined substrate deposition method and the inkjet printing are dealt with in more detail. Both production techniques are analysed in terms of their environmental impacts in this study.

3.1.1 Layers and Materials of a Coated Conductor Produced by Inclined Substrate Deposition

3.1.1.1 Hastelloy® C-276 Substrate

The substrate for the inclined substrate deposition process is Hastelloy® C-276, which has a width of 12 mm and a thickness of 100 μm. Hastelloy® C-276 is a nickel-containing alloy developed by Haynes International. The most important properties of this alloy are its resistance to oxidation as well as corrosion by acids. During the production of high-temperature superconductor tapes, the substrate must withstand high temperatures and high tensile stress, among other things. These are the properties of Hastelloy® C-276 that have established it as the substrate for most high-temperature superconductor production processes [110]. Hastelloy® C-276 consists of approximately 57 % nickel. Other ingredients are chromium (~ 16 %), molybdenum (~ 16 %), iron (~ 5 %), tungsten (~ 4 %), cobalt (< 2.5 %), and in minor amounts manganese, silicon, and carbon (< 1 % each).

Hastelloy® C-276 is produced by melting all the metals in an electric furnace and process the melted metalls into an ingot. This ingot is then refined via electroslag remelting. The ingot is then first hot rolled at about 1100 °C to produce sheets. Strips, such as those used for a high-temperature superconductor as substrate, then require additional cold rolling. Detailed information on production in terms of material and energy flows is not known.

The high nickel content of Hastelloy® C-276 must be critically considered, as contact with nickel can have a variety of effects on human health [111]. The environmental impact of nickel production is highly dependent on where the nickel is produced. Input materials and the metallurgical processes carried out in China differ greatly from those in the rest of the world [112].

Hastelloy® C-276 is well suited as a substrate when the high-temperature superconductor tape is to be used in cables, transformers, or high field solenoids. When used in superconducting fault current limiters, sapphire (Al_2O_3) is an alternative substrate, as it is a dielectric substrate with higher thermal conductivity as well as higher diffusivity [113]. This allows higher electric fields to be achieved. However, compared with Hastelloy® C-276, sapphire is significantly more expensive as a substrate [114]. Other alternatives to Hastelloy® C-276 include CrNiMo stainless steel, which is less expensive but limited by its mechanical properties [115].

3.1.1.2 Magnesium Oxide Buffer Layer

Magnesium oxide is a magnesium salt that is used in a wide variety of applications, such as the production of cement or in clinical applications [116] [117]. Magnesium itself is an abundant element, occurring both in the earth's crust and in seawater [117]. Worldwide reserves for MgO are about 7.6 billion tons, and the resources are virtually unlimited [118]. The largest producer of MgO is China with about 18 million tons in 2020, followed by Brazil (1.5 million tons), Russia (1.5), and Turkey (1.1) [118]. Nevertheless, according to the critical raw materials report by the European Commission, magnesium is one of the most critical materials due to its high economic importance. Furthermore, Magnesium poses a high supply risk as it is currently mainly extracted in China [119].

In a rare earth barium copper oxide superconductor, MgO is used as a buffer layer between the substrate and the superconducting layer. Here, MgO provides the basis for the epitaxial growth of the rare earth barium copper oxide layer. The MgO layer is deposited in two steps by inclined substrate deposition and has a thickness of 3.5 μm.

In contrast to the yttria-stabilized zirconia used in the ion-beam-assisted deposition process, the inclined substrate deposition process uses magnesium oxide for the buffer layer [120]. The advantage of using MgO results in greater time efficiency compared to the ion-beam-assisted deposition process [120].

3.1.1.3 Gadolinium Barium Copper Oxide Superconducting Layer

The group of rare earth elements includes the 15 elements of the lanthanide group as well as yttrium and scandium [121]. These are metals that all have similar geochemical properties and occur together naturally in the same minerals [122] [123]. In addition, all these metals have in common that they have a high electrical conductivity.

Rare earths are used in a wide range of technologies and are often difficult or impossible to substitute [124]. In addition to their use in high-temperature superconductors, they are also used among others in metal alloys, vehicle batteries, glass, permanent magnets, or lasers [124] [125]. Because of their wide-spread applications, the global production of rare earths has nearly doubled since 1990 [126].

However, while the name suggests otherwise, rare earths are not rare [127]. Even the scarcest rare earths, namely thulium and lutetium, are more abundant in the Earth's crust than gold or platinum [128] [129]. Nevertheless, rare earths are subject to a certain criticality. This is because rare earth elements do not occur as individual metals, such as gold or copper, but in over 250 mineral structures [130]. Of these minerals, only a few are economically viable to mine, namely bastnaesite, monazite, loparite and the ion-adsorption clays, as 95 % of all rare earth elements occur in only these minerals [126]. Their global deposits, in turn, are limited to a few locations. The global reserves of rare earths are estimated at about 120 billion tons. The majority of these are distributed among China (44 million tons), Vietnam (22), Brazil (21) and Russia (12) [118]. With a share of about 86 %, China is the largest producer of rare earths. The EU, for example, is completely dependent on imports for rare earths and obtains over 98 % of imported rare earths from China [131]. Due to the economic importance as well as the existing supply risk, the rare earth elements are therefore classified by the EU as a critical raw material [119].

In addition to locally limited deposits, Chinese trade policies contribute to the criticality of rare earths. For instance, China has introduced export quotas and additionally exports only rare earths that have already been domestically processed, in the form of metals, alloys or oxides [123] [124]. Furthermore, the mining and refining processes cause environmental damage through, for example, emissions of fluorine and radioactive thorium [126] [132]. This has resulted in increased global interest in the recycling of rare earth elements [132]. A major challenge here is that rare earths are usually used only in small quantities [130]. The same applies to the use in rare earth barium copper oxide superconductors. In the case of a $GdBa_2Cu_3O_{7-\gamma}$ conductor, as it is used in the examined inclined substrate deposition high-temperature superconductor tape, Gadolinium accounts for just 21.4 % of the molar mass of the superconducting layer. However, the superconducting layer only accounts for a small share of the total tape thickness (about 3 %). Thus, it is virtually impossible to extract the rare earth from the tape during recycling as its share in the total mass of the entire tape is negligible. Therefore, the rare earth gets lost during the end-of-life treatment where the tapes is melted down as steel scrap.

3.1.1.4 Silver Metallisation

Silver is used in a wide variety of applications due to its diverse properties. Due to antibacterial properties, it is used in medical equipment, while the property of being the best metallic conductor also makes it widely used in electrical engineering. Furthermore, silver is used in jewelry and as currency. Globally, about 25,000 tons of silver were produced in 2020, with the majority coming from Mexico (5,600 tons) [118]. Regarding the criticality of silver, it is assumed that the supply of silver will be at risk as early as 2075 and that from 2100 silver will have to be sourced mainly from recycling and urban mining [133]. The EU does not yet classify silver as a critical element but increased the supply risk indicator from 0.5 in 2017 to 0.7 in 2020 [131] [119]. From a value of 1.0, silver would be classified as a critical element due to its high economic importance.

In the superconductor, silver is used both as protection and as an electrical contact. During a quench, the superconductor becomes resistive. In this case, the silver layer becomes the current carrying layer and protects the tape from damage [134]. Depending on the architecture, silver can be applied either on one side of the rare earth barium copper oxide layer or as a surround layer around the entire tape. In both cases, the layer has a thickness of about 1-2 μm.

3.1.1.5 Copper Shunt

In certain cases, a tape may also have a shunt. This also serves the electrical stability and mechanical robustness. In the architecture studied, this shunt consists of copper, which has a high electrical conductivity. The shunt is applied on one side of the tape and has a thickness of 40 μm. However, depending on the intended application the shunt can also be thicker (about 100 μm) or thinner (10 μm) surrounding the tape on all sides.

In 2020, about 25 million tons of copper were produced worldwide, of which the largest part came from China with almost 10 million tons. Global copper reserves are 870 million tons, while identified resources are about 2.1 billion tons [118]. Copper is therefore not currently assessed as a critical element by the European Commission [119] [131] [135].

As an alternative to copper, Hastelloy® C-276 alloy can also be used for the shunt. A 500 μm thick Hastelloy® C-276 shunt was used for example for the high-temperature superconductor tapes produced in the FastGrid project, which are to be used in a fault current limiter [136].

3.1.2 Layers and Materials of a Coated Conductor Produced by Inkjet Printing

3.1.2.1 Sapphire Substrate

For the inkjet printing process, sapphire substrate (Al_2O_3) is used, which has a thickness of 500-1000 μm. Sapphire is very suitable for use in superconducting fault current limiters because of its high thermal conductivity, low dielectric constant, and mechanical strength [137]. While the substrate provides the texture, the lattice is not yet suitable for the the growth of the superconducting layer. Therefore, an additional buffer layer is necessary to influence the texture quality of the rare earth barium copper oxide layer.

Alumina is an abundant element in the earth's crust, which is mainly extracted from bauxite [138]. In 2020, about 136 million tons of aluminum oxide (also called alumina) were produced worldwide, with most of this coming from China (74 million tons) and Australia (21 million tons). While the global bauxite reserves are 30 billion tons, the resources are estimated to be up to 75 billion tons [118]. The European Commission does not currently classify aluminum as a critical element. Although the economic importance is assessed as relatively high, no supply risk is assumed [119].

3.1.2.2 Ceria-Zirconia Buffer Layer

Since the substrate itself does not have sufficient texturing, a buffer layer is printed onto the substrate. The ink for this layer consists of a ceria-zirconia ($Ce_{0.9}Zr_{0.1}O_2$) propionic based ink and the printed layer has a thickness of about 20-30 nm.

Often, either yttria-stabilized zirconia or cerium dioxide CeO_2 is used for a buffer layer [139]. However, CeO_2 has a critical thickness of 50 nm and is prone to microcracking. These microcracks can cause a reaction between the substrate and the rare earth barium copper oxide layer, which decreases the critical current density [140]. To counteract this effect, the cerium dioxide layer is doped with zirconium to reduce the microcracks caused by a lattice mismatch between the substrate and the buffer layer [140] [141] [142].

Cerium itself belongs to the light rare earth elements and is classified as critical by the European Commission, as are the other rare earth elements [119]. It is subject to the same market conditions that have already been described in chapter 3.1.1.3.

Worldwide, about 1.4 million tons of zirconium were produced in 2020 [118]. The main producers are Australia (480 thousand tons) and South Africa (320 thousand tons). Global reserves are about 64 million tons, but zirconium is not classified as a critical element by the European Commission, which is also due to the fact that there is a suitable substitute for zirconium for many applications [118] [119].

3.1.2.3 Yttrium Barium Copper Oxide Superconducting Layer

In inkjet printing, which is investigated in this study, a rare earth barium copper oxide layer is printed, which is about 300 nm thick and uses yttrium as a rare earth element. Yttrium is the most commonly used rare earth element [124]. In general, however, mining and market conditions do not differ from those of other rare earths as described in chapter 3.1.1.3. As with the gadolinium barium copper oxide ceramic, yttrium accounts for only a small fraction of the molar mass of the superconductor layer (about 14 %). This makes targeted recycling impossible as the superconducting layer only accounts for a fraction of the entire tape thickness. There are also no suitable substitutes for yttrium in many applications [124]. In the case of superconductors, only the use of yttrium in yttria-stabilised zirconia buffer layer can be substituted by other elements, such as magnesium oxide. In the actual superconducting rare earth barium copper oxide layer, only another rare earth element such as gadolinium or dysprosium can be used.

3.1.2.4 Current Flow Diverter

In the investigated architecture, a so-called current flow diverter is printed as on top of the yttrium barium copper oxide layer. This layer has a thickness of about 100 nm and is printed from an ink consisting of yttrium acetate, butanol, diethanolamine and propionic acid.

The purpose of this layer is to protect the tape from defects by increasing the normal zone propagation velocity [108]. The normal zone propagation velocity is the speed at which a hot spot, which is a thermal instability, moves along the conductor. The name comes from the transition of the conductor from the superconducting to the normal, non-superconducting state. If the normal zone propagation velocity is low, local damage to the conductor may occur, which the current flow diverter is designed to reduce. The current flow diverter is a resistive layer on the superconductor that forces the current along a specific path at the edge of the tape, creating only a partial quench along the conductor cross-section [108]. A stabiliser layer, made of copper for example, can also be applied to the current flow diverter, but this was not done in the architecture investigated in this study.

3.2 Goal and Scope Definition

3.2.1 Goal

The goal of this study is to analyse the environmental impacts of the production of high-temperature superconductors and to identify the environmentally most crucial steps within the production chain. Since high-temperature superconductor are still a relatively new technology, the production processes are still at a stage that allows further optimization. In addition, novel production processes will continue to be investigated. For this reason, two different production processes are examined in this study regarding their environmental impact. The inclined substrate deposition process, as performed by the company THEVA, is already in commercial use. In contrast to this is the production with an inkjet printer, as carried out by the company Oxolutia. This process is in its early stages and is only carried out on a laboratory scale.

In such an attributional LCA, primary data are needed from manufacturers regarding all their production processes, as literature data or aggregated process data may not be able to cover the differences of the individual production routes with the necessary level of detail. However, it also follows that the results of this study cannot necessarily be applied to other types of production of high-temperature superconductor tapes that were not considered.

The environmental impact of the production of high-temperature superconductor tapes should be as detailed and broad as possible. Therefore, the impact assessment methods Environmental Footprint 3.0 and cumulative energy demand mentioned in chapter 2.4.2 are used. In the case of the Environmental Footprint 3.0 method, all impact categories mentioned are considered.

The aim of this study is to find out which production steps have the greatest environmental impact and to identify any potential for optimisation. The target group of this study are therefore manufacturers of high-temperature superconductor tapes who want to make their production more environmentally friendly. In addition, this study should also serve as a source of information on the environmental impact of the high-temperature superconductor tapes used by manufacturers of technologies that use high-temperature superconductor tapes.

An additional aim of this study is to conduct a prospective LCA to analyse potential future developments in the production of high-temperature superconductors. As prospective LCA cover potential future developments they inherently introduce uncertainty to the model. This uncertainty concerns the data of the model as well as potential future circumstances such as changes in electricity mixes. In this study, to address this concern potential future production scenarios are developed in cooperation with the tape manufacturers. These scenarios include production techniques with higher material and energy efficiency as well as future developments of the tape architecture itself.

3.2.2 Scope

In this attributional LCA, two product systems are examined regarding their environmental impacts. The two product systems have the same function, namely the production of 2G high-temperature superconductor tapes. The function can be quantified by the amount of superconductor produced, which is why the functional unit in this study is defined as one metre of high-temperature superconductor tape produced. However, it must also be considered that the current carrying capacity of the produced tapes varies depending on the production route. The high-temperature superconductor tape from THEVA has a current carrying capacity of about 600 A at a width of 12 mm. This is significantly more than the high-temperature superconductor tape from Oxolutia, which comes to 264 A at a width of 12 mm. For a direct comparison, it is therefore necessary to additionally standardise the results based on the current-carrying capacity.

The product systems are considered in the so-called cradle-to-gate approach. This means that of the entire life cycle of the high-temperature superconductor tapes, only the part up to production is considered. The use of the high-temperature superconductor tapes is outside the scope of this LCA. Thus, the product systems include all production steps that take place at the production site and which can be influenced and changed by the manufacturer. The inventory data for these processes have to be based on primary data from the production site itself, if possible, to ensure satisfactory quality of the results. Figure 3.5 shows the system boundaries of the product system that represents the inclined substrate deposition process of THEVA. This product system had at the time of the investigation about a 60 % yield of functional tape.

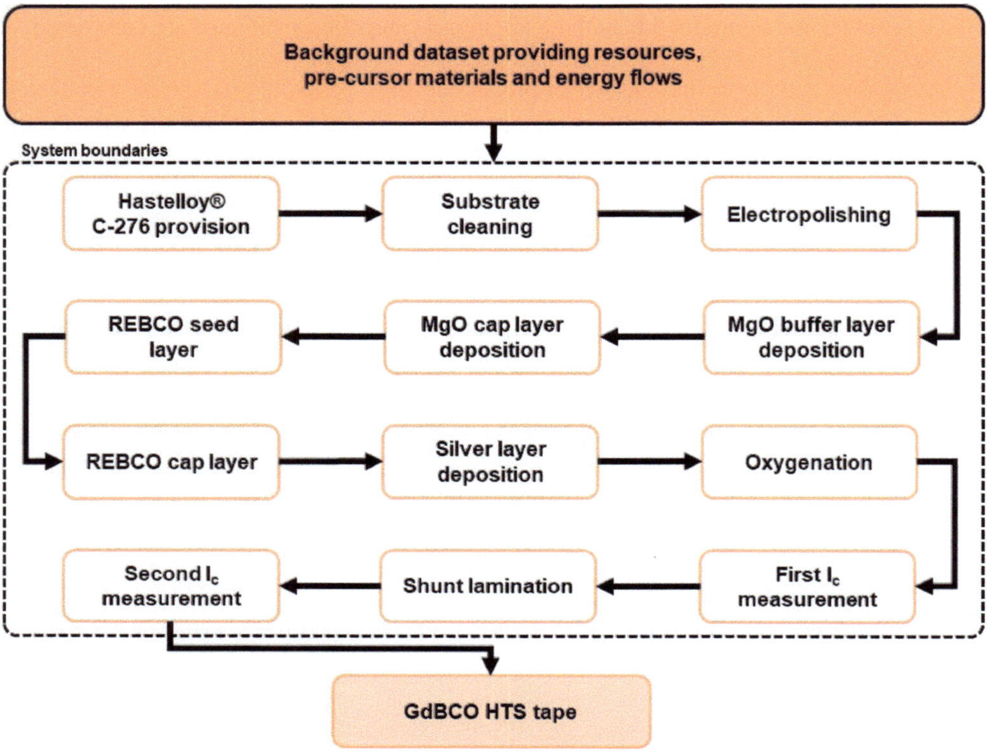

Figure 3.5: System boundaries of the THEVA inclined substrate deposition process.

The system boundaries of the Oxolutia inkjet-printing product system are shown in Figure 3.6. Since this product system is only on a laboratory scale, the yield of functioning high-temperature superconductor tape here is assumed to be only about 10 %.

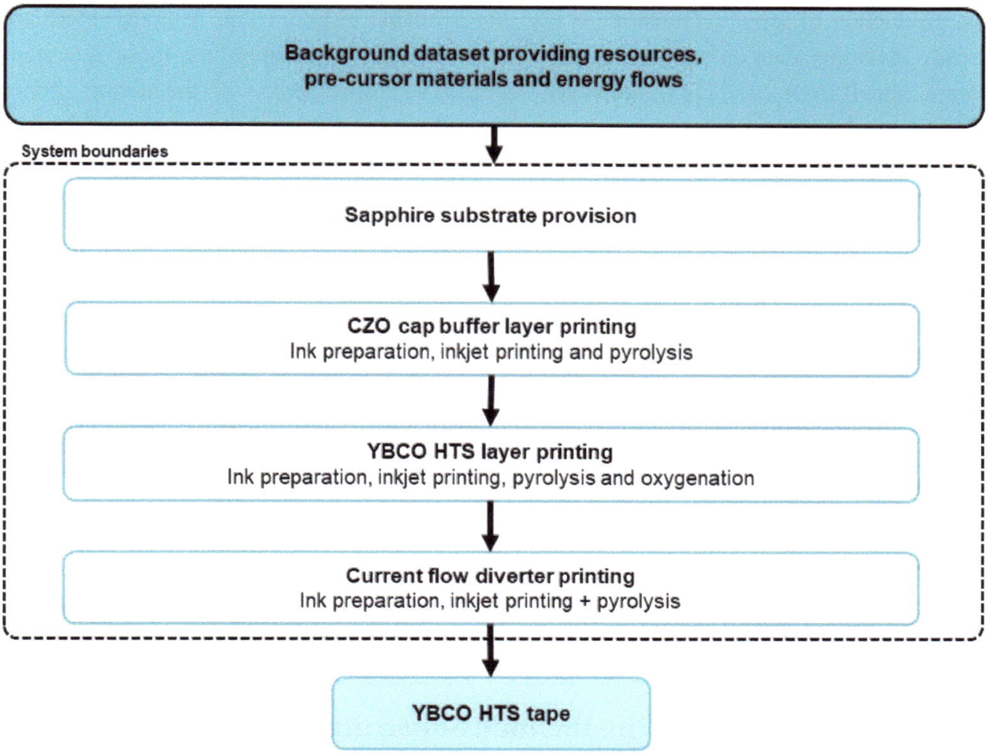

Figure 3.6: System boundaries of the Oxolutia inkjet-printing process.

A background database is used for any precursor products or energy flows. In this study, Ecoinvent version 3.5 with the cut-off system is used for this purpose [86]. This leads to a few limitations of this study. Ecoinvent does not have detailed inventory data for the individual rare earths, but only an aggregated rare earth oxide inventory for most rare earths. However, because most rare earths occur in compounds anyway, this limitation should not have a too great impact, as the mining and refining processes are the same in all cases.

Both product systems have limitations regarding the substrates. The tape produced by THEVA use Hastelloy® C-276 as substrate. This alloy is not available in Ecoinvent and therefore an own modell of the production has to be created. The literature does not contain any detailed data on the energy and material flows of production, and different manufacturers were also unwilling to provide precise information on this. Therefore, only the material composition and one standard hot and one standard cold rolling process from Ecoinvent were inserted to approximate the environmental impacts of Hastelloy® C-276 production. Furthermore, it was assumed that the alloy contains an average share of recycled metals. The average recycled content values for the individual metals were taken from the United Nations Environment Programme [53].

The production of sapphire substrate is also not available in Ecoinvent. Although there are already inventory data on sapphire production to be found in the literature, these data stem from a modell from 2011 [143]. However, based on correspondence with the authors of this study, it was decided not to use this data as it was collected at a time when sapphire production was changing significantly quite fast and this data would no longer be representative today. Thus, only the amount of Al_2O_3 material used is considered.

To make the results more robust, an uncertainty analysis in the form of a Monte Carlo simulation is carried out for both product systems. In addition, a sensitivity analysis is carried out for both product systems by analysing the influence of a layer thickness reduction for individual layers.

Furthermore, scenario analyses are carried out for both product systems. In the case of Oxolutia tape, a theoretical upscaling from laboratory scale to commercial scale is carried out by increasing the yield to the same value as the THEVA tape. For the tape from THEVA, planned changes in production that result in increased material and energy efficiency are examined.

3.3 Life Cycle Inventory Analysis

3.3.1 Process Chain of the Inclined Substrate Deposition

The high-temperature superconductor tape from THEVA is produced using the inclined substrate deposition method. The substrate is vapour-deposited with the individual layers over several stations. However, the substrate itself is held at a special angle to influence the growth direction of the crystals so that the high-temperature superconductor tape has the highest possible current-carrying capacity in self-field.

The individual production steps and their inventory data are explained in more detail below and are shown in Figure 3.7. The exact quantities of the individual flows are not published here due to a confidentiality agreement with the manufacturer. However, the inventory data was collected directly on site during several visits to the production facility and corrected and refined over several iterations in cooperation with the manufacturer. The data therefore has the highest possible data quality, as it comes from the production site under investigation and relates to the current production processes. The environmental impacts per kg of the five main materials, based on the Environmental Footprint 3.0 impact assessment method and the cumulative energy demand, are shown in Table 3.1. The environmental impacts of all the other input and output flows of the product system are given in appendix B.

The THEVA tape is 12 mm wide and uses Hastelloy® C-276 as substrate. The substrate itself is 100 µm thick. The inventory for substrate production consists only of the materials used in the alloy and two transformation processes. The quantities of the individual metals result from the final composition of the alloy. For each individual metal, a proportion of secondary, recycled metal was assumed in addition to primary metals. The proportions are based on the global average values for recycled content of the individual metals [53]. The two transformation processes were developed based on correspondence with the alloy producer Special Metals Corporation. They are intended to approximate the energy input of production. Due to a lack of information regarding production, no losses during the individual process steps were considered.

The substrate is delivered and cleaned in the first step. For this purpose, the substrate is passed through a water bath. Approximately 100 ml of water is used for cleaning per meter of tape. The water must then be treated afterwards.

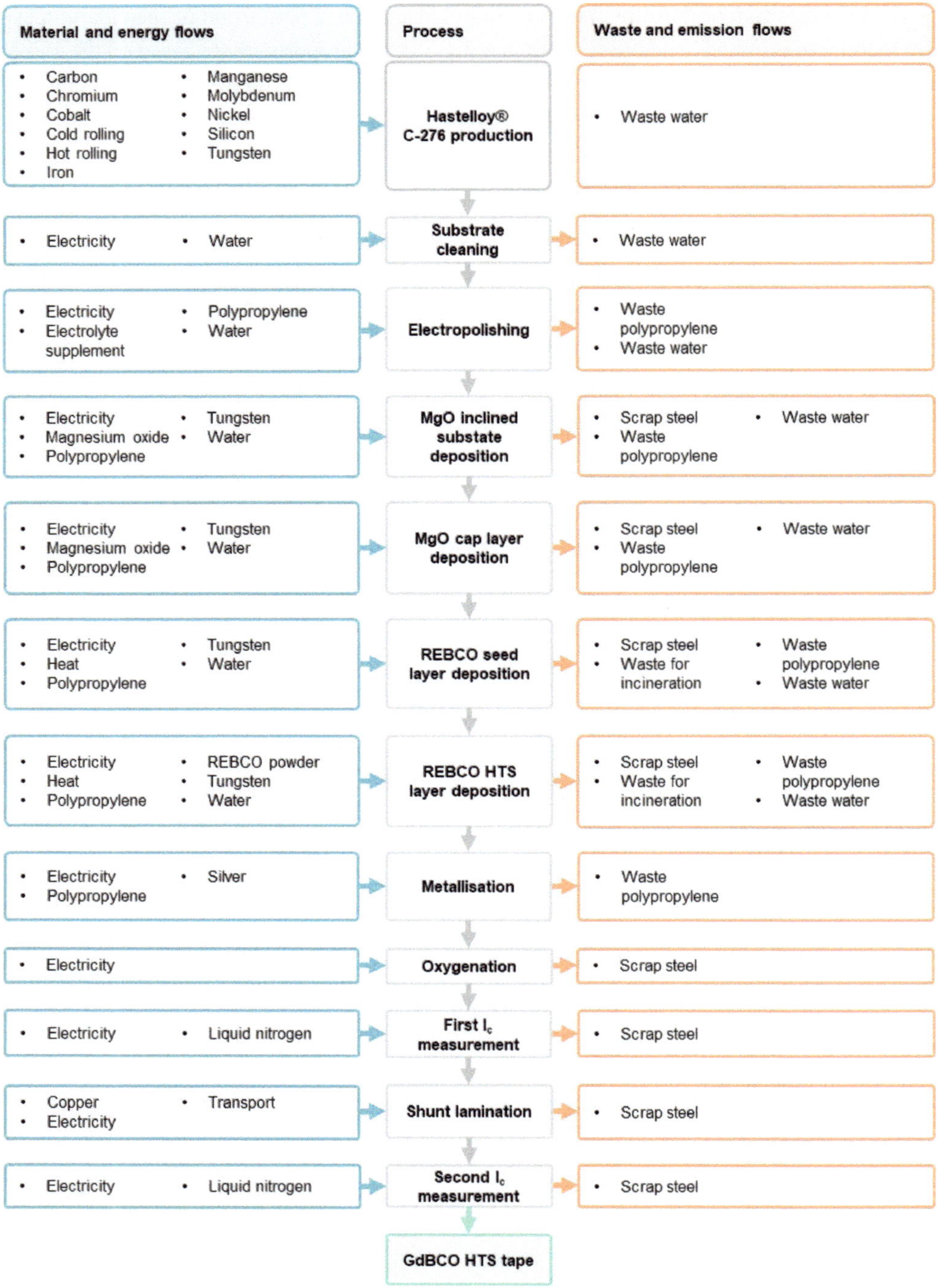

Figure 3.7: Process chain and inventory data of THEVA inclined substrate deposition production. The individual processes (gray) are listed in chronological order. On the left side all material and energy input flows (blue) are listed. On the right are the waste and emmisions flows (orange).

Table 3.1: Environmental impacts of the five main materials used within the inclined substrate deposition process based on the Environmental Footprint 3.0 impact assessment method and the cumulative energy demand.

Impact category	Unit	Copper 1 kg	Hastelloy® C-276 1 kg	Magnesium oxide 1 kg	REBCO powder 1 kg	Silver 1 kg
Acidification terrestrial and freshwater	mol H+ eq.	0.01	2.06	3.03E-03	0.22	0.02
Cancer human health effects	CTUh	7.75E-09	3.20E-06	2.09E-07	2.13E-07	1.36E-08
Climate change	kg CO_2eq.	0.48	18.43	1.19	6.07	1.62
Ecotoxicity freshwater	CTUe	0.41	89.70	23.82	81.70	2.34
Eutrophication freshwater	kg Peq.	1.51E-03	0.49	1.10E-04	0.04	0.01
Eutrophication marine	kg Neq.	6.30E-04	0.10	6.90E-04	0.02	0.01
Eutrophication terrestrial	mol Neq.	0.01	1.15	0.01	0.16	0.07
Ionising radiation	kBq U-235eq.	0.09	1.61	0.03	0.53	0.21
Land use	Pt.	0.84	221.44	0.99	29.41	24.30
Non-cancer human health effects	CTUh	1.66E-07	1.18E-05	1.04E-06	1.64E-05	6.83E-07
Ozone depletion	kg CFC-11eq.	1.93E-08	1.17E-06	2.37E-08	9.91E-07	1.23E-07
Photochemical ozone formation	kg NMVOCeq.	1.26E-03	0.33	2.24E-03	0.05	0.01
Resource use, energy carriers	MJ	6.48	207.07	3.47	99.15	20.61
Resource use, mineral and metals	kg Sbeq.	1.59E-05	0.01	1.94E-07	8.70E-04	2.03E-03
Respiratory inorganics	Disease incidences.	2.61E-08	3.94E-06	9.45E-08	5.87E-07	1.26E-07
Water scarcity	m^3 deprived	107.37	25044.65	59.04	2078.82	380.33
Cumulative energy demand	kWh	16.02	78.19	1.09	31.66	1387.01

For the electricity consumption, the German electricity mix provided by ecoin-vent is considered in each process of the production chain. The electricity mix in ecoinvent is based on data from the International Energy Agency and con-sists of about 70 % fossil fuels, 22 % renewable energy sources and 8 % im-ported electricity [144]. The shares are taken from the year 2014 and are ex-trapolated to be valid for the year 2017 by ecoinvent.

The initial cleaning process is followed by electropolishing of the tape. Here, the tape is further cleaned and the surface smoothed with the help of an electrolysis bath. Just under 10 g of an electrolyte additive is mixed into the bath. This consists of phosphoric acid (70 %), sulfuric acid (25 %) and citric acid (5 %). After electropolishing, the tape is wound up before being passed to the next process step. For this purpose, a polypropylene tape is also wound on as a protective tape. This is to prevent mechanical damage that could occur if the tape is simply wound onto itself. This protective tape is discarded before the next process, when the tape is unwound again for the inclined substrate deposition process, after which a waste treatment process becomes necessary.

In the subsequent inclined substrate deposition process, the MgO buffer layer is evaporated onto the substrate in two steps (inclined substrate deposition and cap layer). In the first part, a

3 µm thick layer is applied, which serves as a base. The 0.5 µm thick cap layer, which is applied in the second step, is intended to smooth the surface, whereby this layer is applied under higher pressure and temperatures. The MgO layer provides the texture for the epitaxial growth of the rare earth barium copper oxide crystals and thus significantly influences the current carrying capacity of the tapes. The total thickness of the layer is about 3.5 µm and requires about 15 g MgO per meter, with an energy demand of about 1 kWh. Losses occur in this process for two reasons. Firstly, immediately prior to the inclined substrate deposition process, the substrates are cut into smaller pieces, removing unusable substrate. Furthermore, due to the nature of evaporation, uneven growth of the MgO layer can occur, again resulting in unusable tape. The unusable pieces are discarded as metal scrap and must be treated. The losses per meter are less than 5 g. As before, the tape is wound up after the inclined substrate deposition process with the aid of a polypropylene tape.

The rare earth barium copper oxide layer is also applied in two partial steps. The seed layer is about 0.5 µm thick and functions as a transitory layer to enable a high quality deposition of the subsequent superconducting layer. It thus significantly determines the current-carrying capacity of the entire tape, since defects in this layer propagate into the functional layer. This functional rare earth barium copper oxide layer is about 3 µm thick and is applied using an adjusted evaporation rate. Overall, the quality of this layer depends on the stoichiometry of the rare earth barium copper oxide powder, the temperature as well as the evaporation rate. Approximately 6 g of rare earth barium copper oxide powder, consisting of gadolinium, barium and copper oxide, is used for both layers. The energy demand (electricity and heat) throughout the entire supply chain is about 4 kWh. The material losses, which occur during this step amount to about 6 g per meter.

The metallisation layer consists of silver, which is evaporated onto the tape. This layer is about 1.5 µm thick and offers both mechanical protection and protection against external environmental influences. It also serves as an electrical contact layer between the superconductor layer and the shunt. The silver that does not land on the tape during evaporation is collected and evaporated again so that there is no loss of silver. In total, about 4 g of silver per metre are required for this process, with an energy demand of about 0.4 kWh.

The next step in the process chain is oxygenation. Oxygenation is an essential step, as oxygen is loaded into the superconducting layer. Only through this process does $GdBa_2Cu_3O_6$ become the superconducting $GdBa_2Cu_3O_{7-\gamma}$. This process only requires about 0.3 kWh of energy to store the oxygen. However, losses may occur again due to unusable tape.

After oxygenation, a quality test takes place in which the critical current I_c of the tape is measured. Liquid nitrogen is needed to cool the tape to operating temperature. If a tape does not meet the quality requirements, it is considered a loss. Otherwise, it can now already be considered a finished high-temperature superconductor tape, provided no shunt is to be applied.

In this study, however, an architecture with a shunt is analysed, which is why shunt lamination is also considered in the process chain. This shunt is made of copper and is approximately 40 μm thick. In total, a little less than 5 g of copper is needed per metre of tape for the shunt. The energy demand of the lamination amounts to about 0.7 kWh. However, the lamination with the copper shunt is not done at the same production site of the rest of the production chain, so a transport process is necessary.

After lamination with the shunt, a quality test is carried out again. This second I_c test is similar to the first test, which is why both tests are always considered together in the following for reasons of clarity. A total of about 16 g of liquid nitrogen and about 9 Wh of electrical energy is required for both tests.

3.3.2 Process Chain of the Inkjet Printing

Oxolutia produces its high-temperature superconductor tape using inkjet printing. The individual layers are printed on a 0.5-1 mm thick sapphire substrate. The tape has a width of 12 mm. The inventory data of the process steps are explained in more detail below, although no exact quantities can be published. In contrast to the THEVA tape, however, the data sets for the Oxolutia tape were collected only through personal correspondence with the manufacturer. There was no visit to the production site itself.

However, according to the manufacturer, the data has been updated and should therefore reflect the production processes fairly well. Overall, however, the data quality should not be rated quite as high as is the case with the tape from THEVA. This is also because the Oxolutia production process was still in a research and development phase at the time of data collection. At the same time, the individual sub-steps were not described in detail by the manufacturer, but rather collected as aggregated data sets for the individual layers. The process chain of the main processes and all input and output flows are shown in Figure 3.8. The environmental impacts of the six main materials for the different layers are shown in Table 3.2. The environmental impacts of all other input and output flows of the product system are provided in appendix C.

At the beginning of the production chain, the sapphire substrate must be obtained. A total of about 36 g of sapphire substrate is needed for one metre of tape. As mentioned earlier, no contemporary information regarding the production of sapphire substrate was found in the literature. Although manufacturers were also contacted, none were willing to provide data. For this reason, only the mass of sapphire (Al_2O_3) is considered, with the process adopted from Ecoinvent.

The sapphire substrate is now first printed with the cap buffer layer. The ink for this printing process consists of roughly 5 g/m of cerium and 0.4 g/m of zircon. Before printing the layer, the substrate is cleaned manually with isopropanol (> 1 g) and the use of polyethylene wipes which must be discarded in the following, making polyethylene treatment necessary. Print heads, which are mainly made of aluminium, are used for the printing process. One of these print heads weighs about 95 g and can print about 691 metres of tape before it needs to be replaced. The inks are then dried, emitting volatile organic compounds (~ 22 mg/m). Finally, pyrolysis is carried out, during which carbon dioxide (~ 10 mg/m) is emitted. The final layer is a $Ce_{0.9}Zr_{0.1}O_2$ layer. In total, about 1.3 kWh per metre is required for all sub-steps of the printing process, using the Spanish electricity mix.

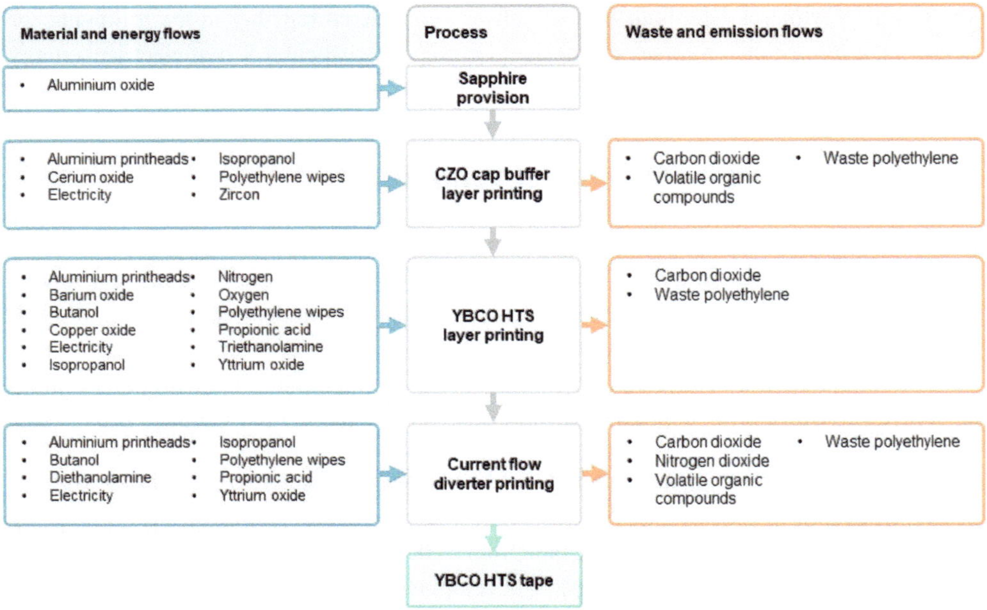

Figure 3.8: Process chain and inventory data of Oxolutia inkjet-printing production. The individual processes (gray) are listed in chronological order. On the left side all material and energy input flows (blue) are listed. On the right are the waste and emmisions flows (orange).

The process steps for printing the superconducting yttrium barium copper oxide layer are similar to those of the ceria-zirconia layer. First, the ink is prepared. This ink contains yttrium (~ 4 mg/m), barium, copper (~ 12 mg/m), the solvents butanol (~ 30 mg/m) and propionic acid (~ 30 mg/m), and triethanolamine (~ 4 mg/m). The underlying ceria-zirconia layer is cleaned again with isopropanol and polyethylene wipes before the actual printing process takes place. After the printing process, the ink is dried and then pyrolysed at around 500 °C using oxygen and nitrogen gas. The last step is oxygenation at around 810 °C, which turns the yttrium barium copper oxide ceramic into a superconducting layer. The total energy consumption amounts to ~ 2.4 kWh/m.

Table 3.2: Environmental impacts of the main materials used within the inkjet printing process based on the Environmental Footprint 3.0 impact assessment method and the cumulative energy demand.

Impact category	Unit	Aluminium oxide 1 kg	Barium oxide 1 kg	Cerium oxide 1 kg	Copper oxide 1 kg	Rare earth concentrate 1 kg	Zircon 1 kg
Acidification terrestrial and freshwater	mol H+ eq.	0.02	0.02	0.03	0.34	0.01	0.01
Cancer human health effects	CTUh	4.22E-07	4.43E-0.8	7.34E-08	2.84E-07	1.33E-08	9.98E-09
Climate change	kg CO_2eq.	1.69	4.00	5.09	3.82	1.56	1.47
Ecotoxicity freshwater	CTUe	5.61	222.03	4.16	42.52	0.96	0.43
Eutrophication freshwater	kg Peq.	7.40E-04	1.59E-03	2.43E-03	0.07	5.30E-04	9.20E-04
Eutrophication marine	kg Neq.	2.43E-03	3.70E-03	4.83E-03	0.03	1.96E-03	2.52E-03
Eutrophication terrestrial	mol Neq.	0.03	0.05	0.05	0.22	0.02	0.03
Ionising radiation	kBq U-235eq.	0.05	0.33	0.45	0.34	0.09	0.10
Land use	Pt.	2.89	6.70	9.67	39.90	6.48	19.03
Non-cancer human health effects	CTUh	9.57E-07	1.45E-05	7.80E-07	2.20E-05	1.80E-07	1.08E-07
Ozone depletion	kg CFC-11eq.	1.31E-07	2.28E-07	1.38E-06	2.47E-07	3.02E-07	1.15E-07
Photochemical ozone formation	kg NMVOCeq.	7.27E-03	0.01	0.02	0.06	5.07E-03	7.08E-03
Resource use, energy carriers	MJ	16.58	39.72	112.16	43.45	19.04	17.56
Resource use, mineral and metals	kg Sbeq.	2.36E-05	2.33E-05	6.23E-05	1.48E-03	9.47E-06	1.91E-06
Respiratory inorganics	Disease incidences.	1.68E-07	1.58E-07	2.70E-07	7.01E-07	1.05E-07	1.23E-07
Water scarcity	m^3 deprived	137.13	723.65	677.30	2.69E+03	239.97	229.47
Cumulative energy demand	kWh	4.99	12.71	34.34	15.40	6.20	5.57

Finally, a current flow diverter is printed onto the superconducting layer. The current flow diverter is a highly resistive layer designed to protect the tape from the formation of destructive hot spots during a quench [108]. Here, the basic procedure is again the same. First, the ink is prepared. This consists of roughly 6 mg/m yttrium acetate ($C_6H_9O_6Y$), butanol (~ 80 mg/m), propionic acid (~ 40 mg/m) and diethanolamine (~ 15 mg/m). Since ecoinvent does not have inventory data for yttrium acetate, the data set for the rare earth oxide was used to map the required amount of yttrium. The printing process takes place after cleaning the underlying layer. After the ink has dried, pyrolysis takes place again. This consists of three sub-sections, which take different lengths of time and use different temperatures. The total energy consumption amounts up to about 0.7 kWh/m. The result is a Y_2O_3 current flow diverter layer.

3.4 Life Cycle Impact Assessment

3.4.1 Inclined Substrate Deposition

3.4.1.1 Contribution Analysis

Based on the system boundaries presented in chapter 3.2.2, the environmental impacts of the individual process steps are now calculated. The input and output flows shown in chapter 3.3.1 are used for this purpose. As described in chapter 2.4.1, the elementary flows of emissions are assigned to the individual impact categories and converted to an indicator using impact factors. Since this is done individually for each process step, the individual contributions of all process steps can be calculated and presented in the form of a contribution analysis.

Figure 3.9 shows the results of the contribution analysis of the individual layers of THEVA tape. The production steps substrate provision, substrate cleaning and electropolishing are summarised under the item substrate preparation. Likewise, the two magnesium-oxide and the two rare earth barium copper oxide layers have been combined in each case for the purpose of better comprehensibility, so that they each represent a single aggregate layer. The environmental impacts of the two I_c test measurements are also summarised, as they involve the same process twice.

Substrate preparation has an average contribution of 11.6 % across all impact categories. Overall, the contributions vary from 3.2 % in the category resource use energy carriers to 32.4 % in the category cancer human health effects. In the latter category, substrate preparation even has the largest share of all layers. This is mainly due to the chromium used in Hastelloy® C-276, whose production is responsible for 20 % of the total impact in this category.

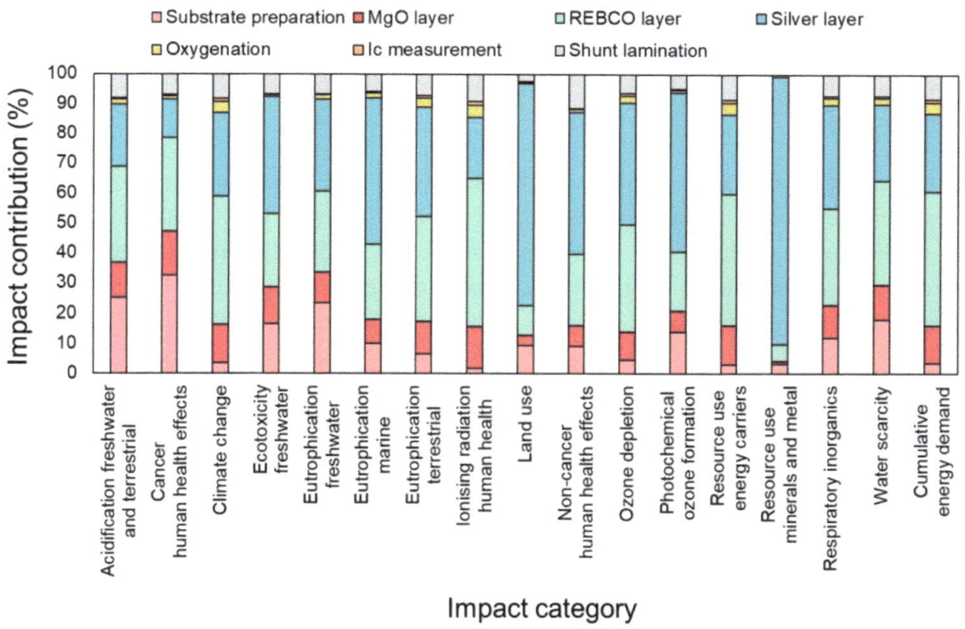

Figure 3.9: Contribution analysis results for the THEVA tape showing the relative impact that each layer has.

The magnesium oxide layers are responsible for about 9.9 % of the environmental impact on average. They make the smallest contribution in the category resource use minerals and metal with just 0.6 %. Their largest contribution is in the category cancer human health effects with 14.7 %. However, this is because cutting processes occur during the production of the MgO layer. These cutting processes result in losses of the underlying substrate. For one metre of MgO layer that can still be processed, more than one metre of substrate must be produced, which is reflected in the effects for the MgO layer.

The rare earth barium copper oxide layer has the largest share of environmental impacts in six categories and are responsible for 30.3 % of impacts on average. They have the lowest share in the category resource use minerals and metal with 5.5 %. However, with 49.1 %, rare earth barium copper oxide layers are responsible for almost half of the impacts in the category ionising radiation human health. This is primarily due to the energy intensity of the process and the imported nuclear share in the electricity mix used. The energy intensity is also reflected in the category climate change, where the rare earth barium copper oxide layers also have the largest share with 42.4 %. Consequently, the cumulative energy demand for the rare earth barium copper oxide layer is also the highest, with a share of 44.5 %.

The silver layer of metallisation has the largest average contribution of 38.7 % across all environmental categories. It has the largest contribution in ten of the 17 categories. Overall, the shares of the silver layer range between 12.8 % and 89.5 %. The silver layer has the largest contribution in the category resource use minerals and metals, which is due to the mining processes of the silver itself. For the same reason, the land use category is also clearly dominated by the silver layer (74.4 %).

The oxygenation process has a low average contribution of 2.0 %. The largest contribution comes from the ionising radiation human health category, where it has a share of 4.2 %. The same applies to the quality tests of the I_c measurements. Across all impact categories, these show an average contribution of only 0.6 %. In none of the categories do the I_c measurements have a contribution greater than 1.3 %.

Lamination with the copper shunt provides on average 6.8 % of the environmental impact. Lamination with the copper shunt provides on average 6.8 % of the environmental impact. The process has the largest contribution in the category non-cancer human health effects with a share of 11.5 %. These are largely due to the copper production itself.

3.4.1.2 Uncertainty Analysis

To analyse the uncertainty of the results, a Monte Carlo simulation with 1000 iterations was carried out. Although it is recommended in literature to carry out several thousand iterations, an analysis of the results shows that so many iterations are not necessary. Figure 3.10 shows how the mean, median, 5th percentile, and 95th percentile of the climate change results changed with each iteration. While the individual values still change significantly with each iteration over the first hundred iterations, the median and the mean hardly show any changes from about 250 iterations onwards. After about 400 iterations, the fluctuations of the 95th percentile are also hardly present. From this it can be concluded that with about 1000 iterations all fluctuations in the results caused by the data uncertainty are already sufficiently covered.

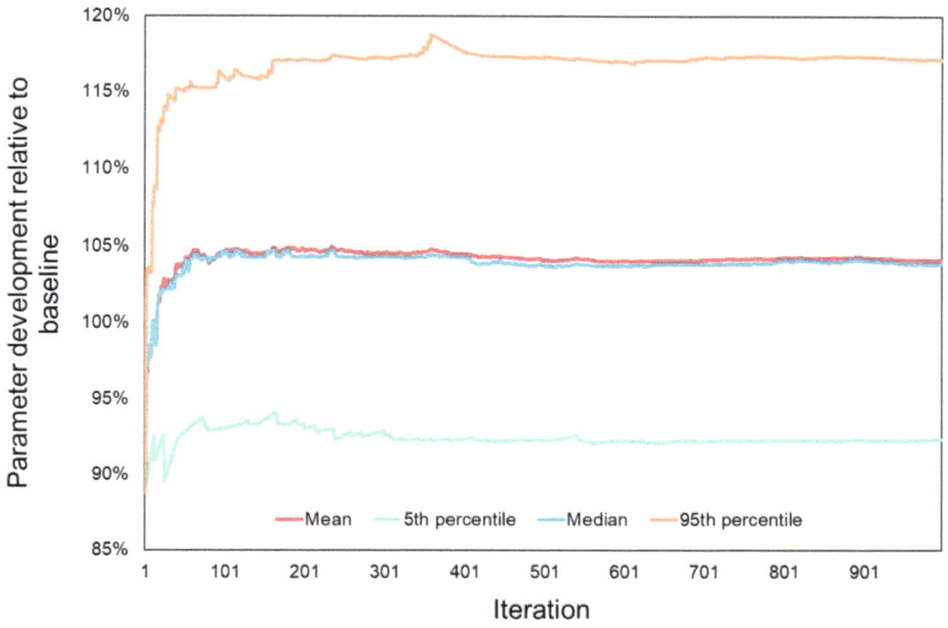

Figure 3.10: Development of the mean, 5th percentile, median, and 95th percentile of the climate change results distribution with each iteration. The 100 % line equals the baseline results from the contribution analysis.

The uncertainties of the individual categories show significant fluctuations, as can be seen in Figure 3.11. In this graph, the 100 % mark of each category represents the corresponding result from the contribution analysis. The boxplots thus show the deviation from this baseline result.

The greatest uncertainty is found in the category eutrophication freshwater, where the range between the 5th percentile and the 95th percentile is over 266 percentage points. The lowest overall uncertainty of less than twelve percentage points is found in the category resource use minerals and metal.

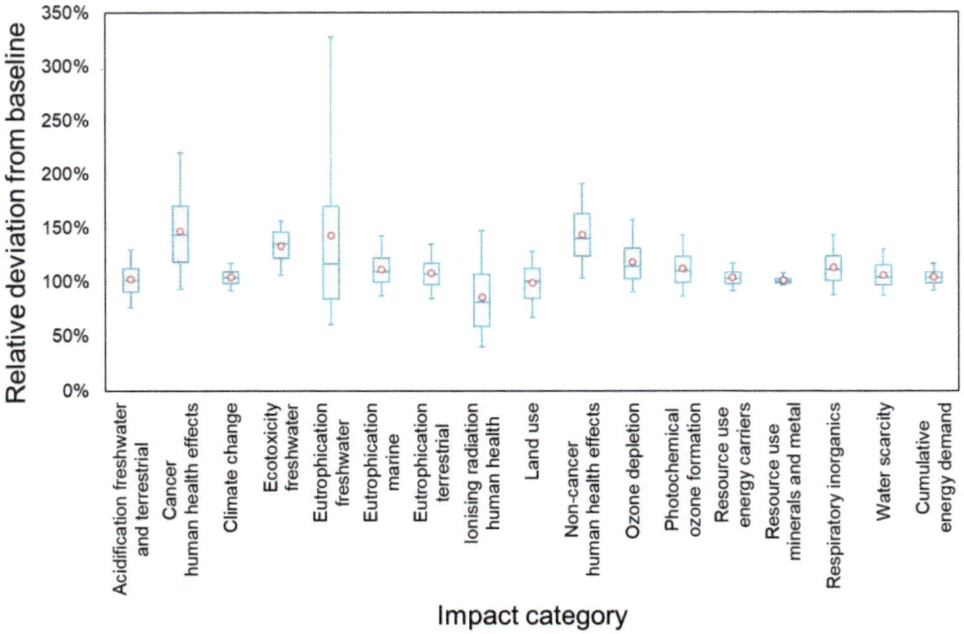

Figure 3.11: Monte-Carlo simulation results of the THEVA tape. The boxes indicate the range from the 1st to the 3rd quartile, while the horizontal line represents the median. The whiskers indicate the margin to the 5th and the 95th percentile. The mean of the result distribution is shown as a red dot. All values are relative with 100 % indicating the baseline result from the contribution analysis.

Four categories have an overall uncertainty of 30 percentage points or less: climate change, resource use energy carriers, resource use minerals and metal, and cumulative energy demand. In these categories, the difference between the 1st and 3rd quartile of results is less than eleven percentage points. Therefore, the results in these particular categories can be assessed as very reliable.

In contrast, however, there are also three categories with an uncertainty of over 100 percentage points: Cancer human health effects, eutrophication freshwater, and ionising radiation human health. The remaining eleven categories vary in their overall uncertainty between 24 and 88 percentage points. On average, the uncertainty margin is 68.5 percentage points.

Overall, it is noticeable that the deviations from the baseline value are significantly greater upwards than downwards. The uncertainty distribution function of the results is therefore not a normal distribution, but rather resembles a log-normal distribution. This was to be expected, as the log-normal distribution is the most common uncertainty distribution function in Ecoinvent. This function is unbounded upwards, while at the same time it cannot take on negative values.

However, it is also noticeable that in the two categories ecotoxicity freshwater and non-cancer human health effects the baseline result is below the 5th percentile. This indicates that the uncertainties in the foreground and background data in these two categories allow significant upward deviations. The baseline results in these categories should therefore even be regarded as outliers. It can therefore be assumed that the environmental impacts in these two categories tend to be higher.

It should be noted at this point that due to the random nature of a Monte Carlo simulation, certain life cycle inventory combinations can occur which are to be considered unrealistic and lead to massive outliers. This is the case, for example, if more steel is required as input in the entire upstream chain of a process than is available as output. In this case, OpenLCA will produce erroneous calculations that are not automatically detected. As there is no automated way to identify which values are mathematically invalid, a thresh-old is established to differentiate between potentially valid and potentially in-valid values. Some of these potentially invalid values can easily be identified as they are more than one order of magnitude larger than the 95th percentile of the Monte Carlo simulation sample, which may be statistically possible but is not probable. However, not every value that is larger than the 95th percentile is automatically invalid and may as well be just an extreme outlier that is still based on mathematically accurate calculations.

Currently, it is not possible to exclude potentially invalid values without being arbitrary to a certain degree. In this study, all values that are at least 50% larger than the 95th percentile are not taken into account. This threshold is assumed to be low enough to exclude all invalid values. At the same time, the threshold is high enough so that extreme values that appear to be based on mathemati-cally valid calculations are still included in the analysis. Nevertheless, due to the arbitrary and subjective nature of this procedure, it is still possible that valid values are removed or that invalid values, which are not as extreme, are still included. In any case, in total less than 10 % of all values are removed due to this procedure so that their influence can be considered as not significant.

3.4.1.3 Sensitivity Analysis

For the sensitivity analysis, the layer thickness of the five layers substrate, MgO layer, rare earth barium copper oxide layer, silver layer, and shunt was reduced by 50 % each. The thicknesses of the other four layers remained the same. The results were then examined to see how they changed in comparison to the baseline result from the contribution analysis. This makes it possible to identify the layer changes to which the overall system reacts most strongly.

Figure 3.12 shows the results of the sensitivity analysis for the categories cancer human health effects, climate change, resource use minerals and metals, and cumulative energy demand as well as for the mean value across all impact categories.

In the category cancer human health effects, a reduction of the substrate thickness leads to an impact reduction of about 16.2 %. The rare earth barium copper oxide layer is on a similar level, where a 50 % reduction of the layer thickness reduces the environmental impact by 15.8 %. The reduction of the shunt thickness has barely any influence, with the environmental impact being reduced by only 3.5 %.

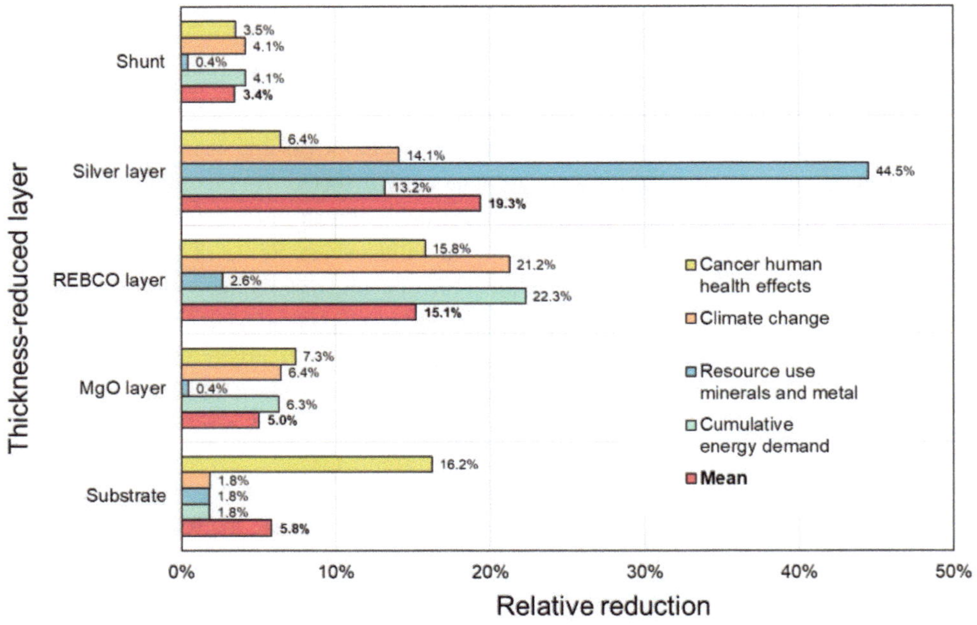

Figure 3.12: Sensitivity analysis results of the categories cancer human health effects, climate change, resource use minerals and metals, and cumulative energydemand. Additionally, the mean of all impact categories is shown. Bars indicate to relative reduction compared to the baseline result of the contribution analysis due to a thickness reduction of the corresponding layer by 50 %.

For climate change, the reduction of the rare earth barium copper oxide layer thickness results in an impact reduction of about 21.2 %. A reduction of the silver layer thickness by 50 % still reduces the environmental impact of the entire system by 14.1 %. A reduction of the substrate thickness, on the other hand, hardly influences the result. Even with a 50 % reduction, the results only change by 1.8 %. This means, that the system reacts more sensitive to a change in layer thickness when the layer already has a high contribution to the climate change category.

A similar conclusion can be drawn for the category resource use minerals and metal. In this category, the silver layer has a share of almost 90 % of the total impact. Consequently, a reduction of the layer thickness by 50 % also significantly reduce the impact in this category. The reduction in this case is 44.5 %. Due to the small share of the other layers in this category, even a significant reduction of the layer thickness by 50 % leads to hardly noticeable changes in the results. For all other layers, the thickness reduction leads to a maximum impact reduction of less than 3 %.

In terms of cumulative energy demand, the result is remarkably similar to that of the climate change category. The greatest influence here is the reduction of the rare earth barium copper oxide layer (22.3 %). The reduction of the silver layer thickness also has a double-digit percentage value (13.2 %). The smallest effect is a reduction in the substrate layer thickness (1.8 %). The similarity to the climate change category is not surprising because greenhouse gas emissions are strongly linked to energy consumption and therefore these two categories often tend to be similar in terms of their results.

Looking at the average sensitivity across all impact categories, a reduction in silver layer thickness has the greatest impact on the overall system (19.3 % impact reduction) and, on average, would have the most positive environmental effect on production. This is interesting in that the silver layer, at 1.5 µm, is the thinnest layer of all. A reduction in the rare earth barium copper oxide layer also results in significant improvements in environmental impact (15.1 %). A reduction of the copper shunt would have the least impact. A reduction of the shunt thickness from 40 µm to 20 µm would reduce the average environmental impact by only 3.4 %.

3.4.1.4 Scenario Analysis

In order to examine a change in the production chain planned by THEVA with regard to its environmental impact, a scenario analysis is carried out. The following changes are planned:

- Increasing material efficiency by abandoning cutting processes
- Increase in material and energy efficiency by combining the process steps for both rare earth barium copper oxide layers
- Increasing material and energy efficiency by adjusting the throughput rates for all layers

The basic architecture of the tape will not change due to the planned changes. For reasons of secrecy, no exact information about the planned changes will be presented here.

Figure 3.13 shows how the planned changes affect the global warming potential of the climate change category. The current production process emits about 5.8 kg CO_2 eq per metre of tape. The planned changes would reduce this value by about 1.9 kg to 3.8 kg CO_2 equivalents. In particular, the changes in the production process of the rare earth barium copper oxide layer are clearly noticeable here.

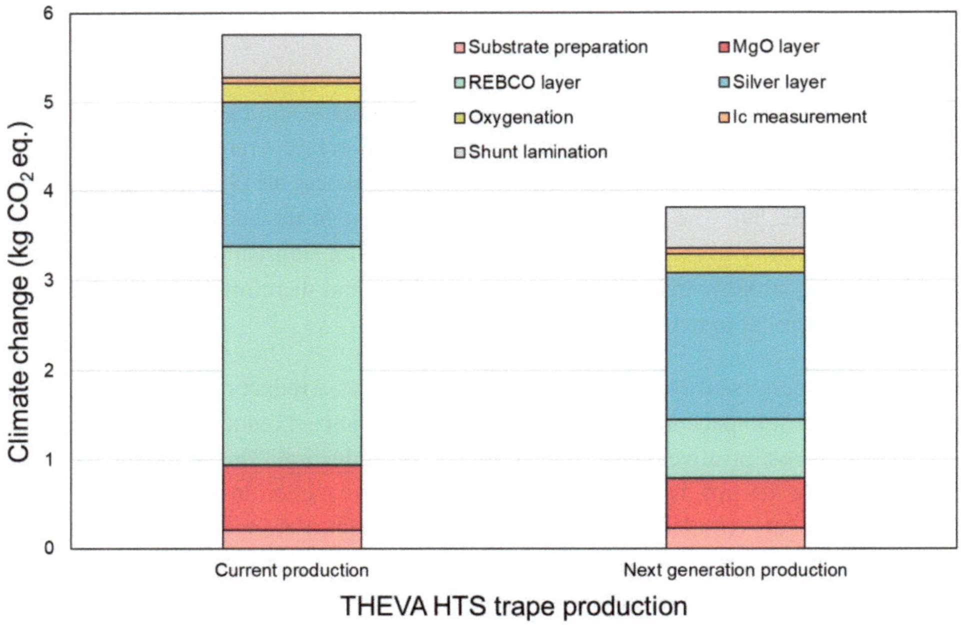

Figure 3.13: Impact comparison of the current production and the planned next generation production in the impact category climate change.

The increase in material efficiency is hardly noticeable in the category resource use minerals and metals, as shown in Figure 3.14. This is because the effects in this category are clearly dominated by the silver layer. However, the demand for silver is only reduced by less than one per cent in next generation production. This change is not noticeable in the environmental impacts. Nevertheless, the impacts drop from 2.3 g Sb eq. to 2.2 g Sb eq. per metre. This is due to the increase in efficiency of the production of the rare earth barium copper oxide layer.

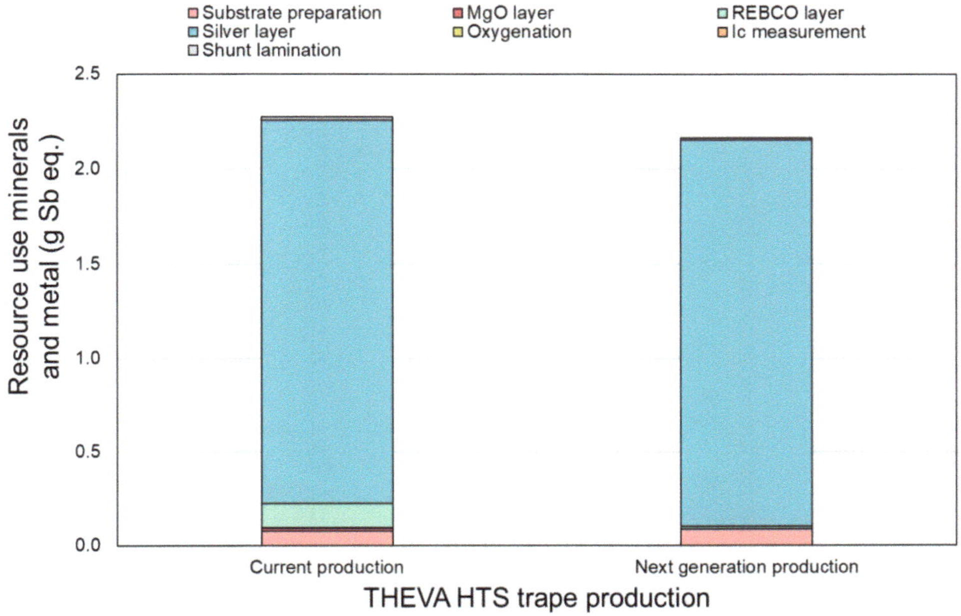

Figure 3.14: Impact comparison of the current production and the planned next generation production in the impact category resource use minerals and metal.

As the planned change also aims to increase energy efficiency, the impact on cumulative energy demand is also considered. Figure 3.15 shows how this is reduced by the next generation production. Current production requires about 25 kWh of energy per metre over the entire part of the life cycle considered. The planned changes reduce the cumulative energy demand by about 9 kWh to 16.5 kWh. This is particularly evident in the reduced energy requirements in the production of the rare earth barium copper oxide layer. Combining the two production steps saves a total of about 8.4 kWh.

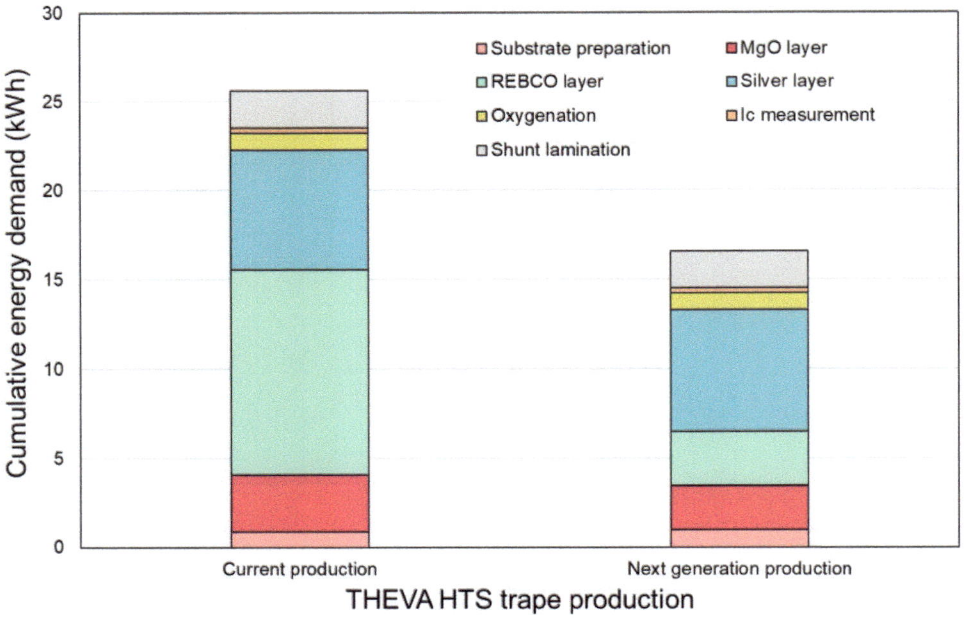

Figure 3.15: Impact comparison of the current production and the planned next generation production in the impact category cumulative energydemand.

3.4.2 Inkjet Printing

3.4.2.1 Contribution Analysis

Figure 3.16 shows the results of the contribution analysis for the inkjet printing process of Oxolutia. Overall, the picture is relatively balanced, with the contributions of the individual layers to the total environmental impact being very similar for almost all categories.

The sapphire substrate has an average share of 9.9 % across all categories. However, in the categories cancer human health effects and resource use minerals and metal, the substrate has the largest share with 48.6 % and 39.6 %, respectively. This is due to the emissions during aluminum oxide production and the provision of this same aluminum oxide. However, except for two categories, the substrate accounts for less than five percent of the total impact in all other categories.

The ceria-zirconia cap layer has an average environmental impact of about 26.6%. Overall, the values vary from 24.8 % to 29.2 %.

The only exceptions here are precisely the two categories in which the sapphire substrate has the largest share. In these, the contribution of the ceria-zirconia layer is 15.3 % and 18.1 %, respectively.

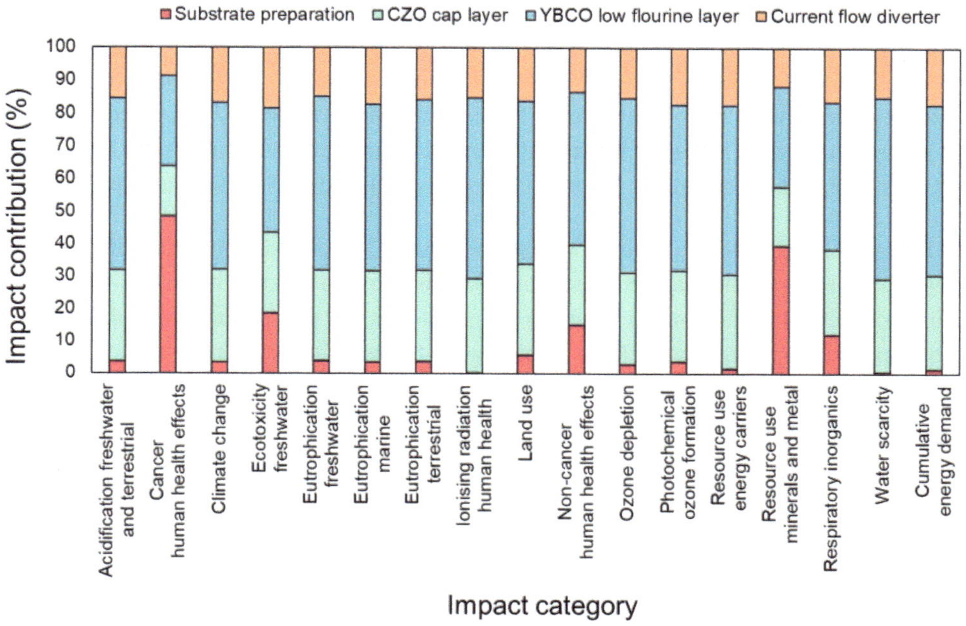

Figure 3.16: Contribution analysis results for the Oxolutia tape showing the relative impact that each layer has.

With the exception of the two categories mentioned above, the yttrium barium copper oxide has the largest share in all other categories, with values ranging from 37.8 % to 55.6 %. This is due to the energy intensity of this process. Electricity consumption emerges as the main source of environmental impact in all categories. In the cancer human health effects and resource use minerals and metal categories, the yttrium barium copper oxide layer has the second largest share of environmental impacts. These are 27.5 % and 30.6 % respectively. On average, the share of the yttrium barium copper oxide layer is 48.0 %.

Printing the current flow diverter has an average contribution to the environmental impact of 15.5 %. The values range from 8.7 % to 18.5 %.

3.4.2.2 Uncertainty Analysis

The uncertainties of the individual categories show significant fluctuations, as can be seen in Figure 3.17. The greatest uncertainty is found in the category cancer human health effects, where the total uncertainty ranges from 71.6 % to 232.9 % (161.3 percentage points). The lowest overall uncertainty of all categories is found in the category cumulative energydemand with an uncertainty of 30.8 percentage points.

Compared to inclined substrate deposition tape, none of the categories for inkjet printing tape has an uncertainty of less than 30 percentage points. The average uncertainty amounts to a total of 63.3 percentage points, which is below the average uncertainty of the inclined substrate deposition tape. This is also because only three categories have an uncertainty in the three-digit range, yet they are not as high as the highest uncertainty of the inclined substrate deposition tape. A total of ten categories have an uncertainty of less than 50 percentage points.

As with the inclined substrate deposition tape, the Monte Carlo simulation results deviate more strongly upwards from the baseline result. Again, this can be attributed to the log-normal uncertainty distribution function that underlies all inventories. As with the inclined substrate deposition tape, the baseline value in the ecotoxicity freshwater category is below the 5th percentile line. Here, the log-normal uncertainty distributions thus increasingly lead to higher values than those assumed in the baseline calculation.

The Monte Carlo simulation results of the inkjet printing tape were additionally corrected for erroneous calculations using the same procedure as described in chapter 3.4.1.2.

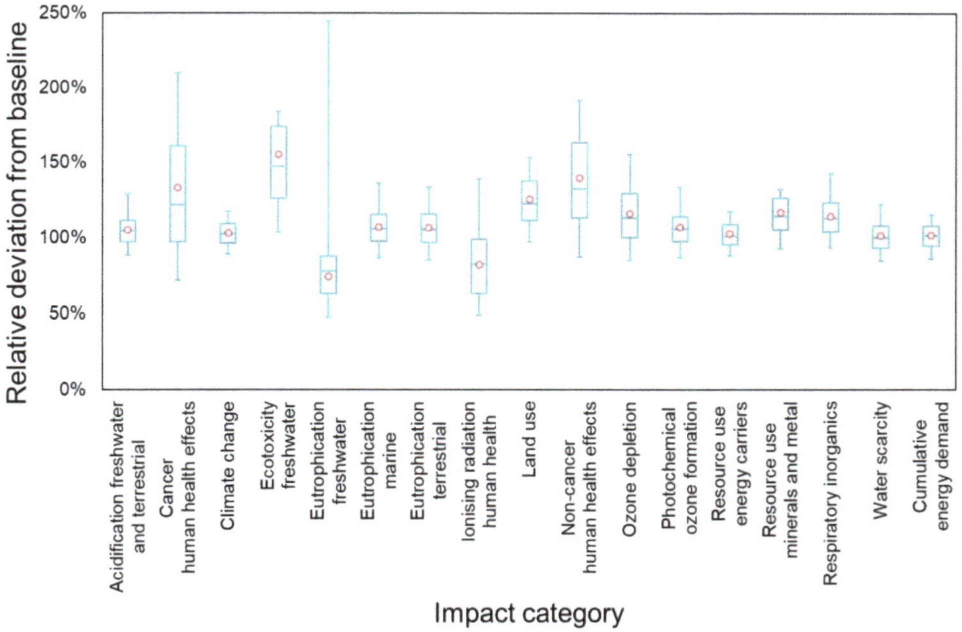

Figure 3.17: Monte-Carlo simulation results of the Oxolutia tape. The boxes indicate the range from the 1st to the 3rd quartile, while the horizontal line represents the median. The whiskers indicate the margin to the 5th and the 95th percentile. The mean of the result distribution is shown as a red dot. All values are relative with 100 % indicating the baseline result from the contribution analysis.

3.4.2.3 Sensitivity Analysis

As with the THEVA tape, a sensitivity analysis was carried out on the Oxolutia tape by reducing the thickness of each layer individually by 50 %. Figure 3.18 shows how these reductions affected the results in the categories cancer human health effect, climate change, resource use minerals and metals, and cumulative energy demand. Additionally, the mean result across all impact categories is shown.

Regarding the category cancer human health effects, the largest environmental savings (24.3 %) are achievable by reducing substrate thickness. A reduction of the yttrium barium copper oxide layer leads to an impact reduction of about 13.7 %. A reduction in the thickness of the current flow diverter has the smallest impact (4.3 %).

The climate change category shows that a reduction of the yttrium barium copper oxide layer thickness reduces the global warming potential by about 25.5 %. A reduction in ceria-zirconia layer thickness leads to a reduction in environmental impact of 14.3 %. With the other two layers, a 50 % reduction in layer thickness reduces the environmental impact by less than 10 %. For the current flow diverter, it is 8.4 % and for the sapphire substrate it is 1.7 %.

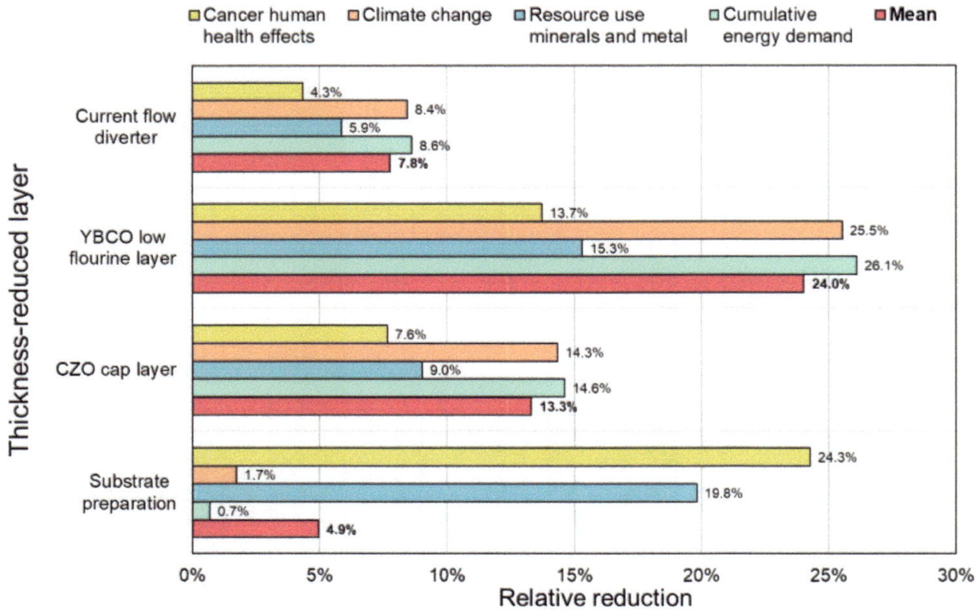

Figure 3.18: Sensitivity analysis results of the categories cancer human health effects, climate change, resource use minerals and metals, and cumulative energydemand. Additionally, the mean of all impact categories is shown. Bars indicate to relative reduction compared to the baseline result of the contribution analysis due to a thickness reduction of the corresponding layer by 50 %.

In the resource use minerals and metals category, the system reacts particularly sensitively to changes in substrate layer thickness. A 50 % change in layer thickness reduces the environmental impact by about 19.8 %. For the yttrium barium copper oxide layer it is as much as 15.3 %, while for the other two layers the impact reductions are less than 10 %.

As with the THEVA tape, the sensitivity of the Oxolutia production system in the cumulative energy demand category is very similar to the sensitivity in the climate change category. A reduction of the yttrium barium copper oxide layer reduces the energy demand by about 26.1 %. The smallest impact in this category is a reduction in substrate thickness (0.7 %).

The average of all impact categories shows that the system reacts particularly to changes in the layer, which on average also contribute the most to the individual categories. Reducing the thickness of the yttrium barium copper oxide layer leads to an average impact reduction of 24.0 %, while the ceria-zirconia layer contributes 13.3 %. The substrate layer has the lowest average sensitivity of 4.9 %.

3.4.2.4 Scenario Analysis

In the scenario analysis for the Oxolutia tape, the yield of the production was increased from 10 % to 60 % to obtain the same yield rate as for the THEVA tape. This is to reflect scaling effects in the transition from laboratory scale to industrial scale. However, as there was no information on how the production process could evolve to achieve such an increase in yield, the material and energy flows of each production step were left the same, resulting in a greater output of usable tape for the same input. Due to the linearity of the LCA method, this procedure leads to predictable results as six times higher yield equals six times lower environmental impact for the initial Oxolutia processes.

However, the original Oxolutia tape architecture does not include a metal stabilisation layer. Without such a stabilisation layer, the tape would be destroyed almost immediately in case a quench occurs. Thus, it is assumed that in the future the Oxolutia tape will also have a silver stabilisation layer. The silver layer production process of THEVA is used to approximate how an additional silver layer affects the environmental impacts of the upscaled Oxolutia tape. On average, the changes in yield and architecture reduced the environmental impacts in 14 impact categories by 69.9 %.

Figure 3.19 shows how this scaling and adding a silver layer affects the impact category climate change. The current process causes greenhouse emissions of 17.9 kg CO_2 eq. over the entire life cycle chain considered. Increasing the yield and adding a silver layer results in an emissions reduction to 4.6 kg CO_2 equivalents. While the current production impacts are well above the values of THEVA production (5.8 kg CO_2 eq.), changes in the production lead to values similar even to THEVA's next generation production (3.8 kg CO_2 eq.).

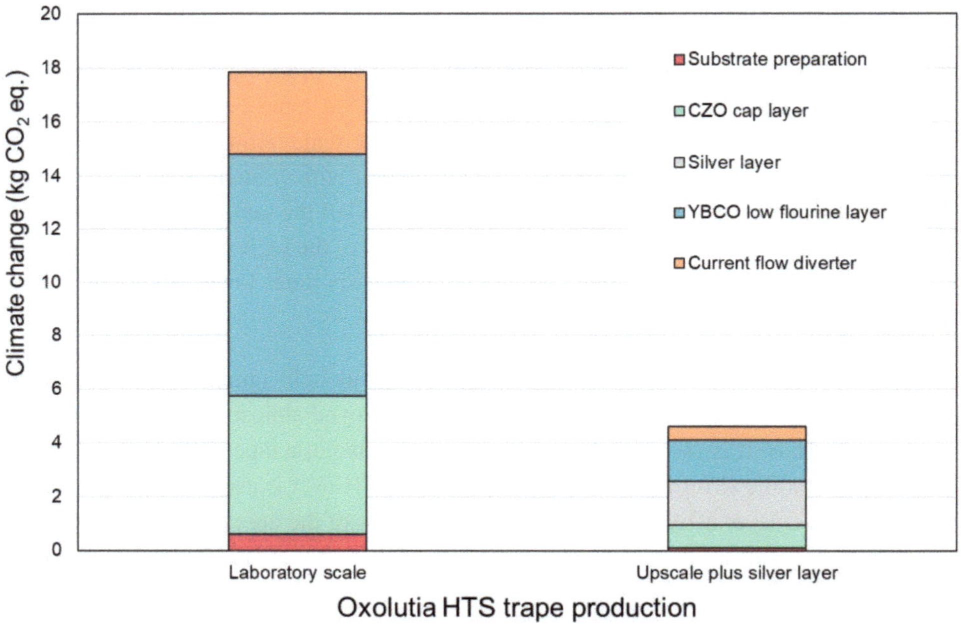

Figure 3.19: Impact comparison of the current yield at laboratory scale and the assumed upscaled yield at industrial scale with an additional silver layer in the impact category climate change.

In the category resource use minerals and metal, the Oxolutia tape already has very low values of just 0.02 g Sb equivalents, as shown in Figure 3.20. The tape from THEVA, with 2.3 g Sb equivalents, is a factor of 1000 higher. However, the laboratory-scaled Oxolutia tape does not have a silver layer, which is the main contributor for the THEVA tape in this impact category. While the yield upscale of the Oxolutia tape does reduce the environmental impacts, the assumed additional silver layer causes the exact opposite effect. The environmental impacts of the Oxolutia tape are significantly increased in this scenario resulting in a resource use of minerals and metal of 2.0 g Sb equivalents. This value is very similar to the value of the next generation production of THEVA (2.1 g Sb equivalents), which can be explained by using the same silver layer production process.

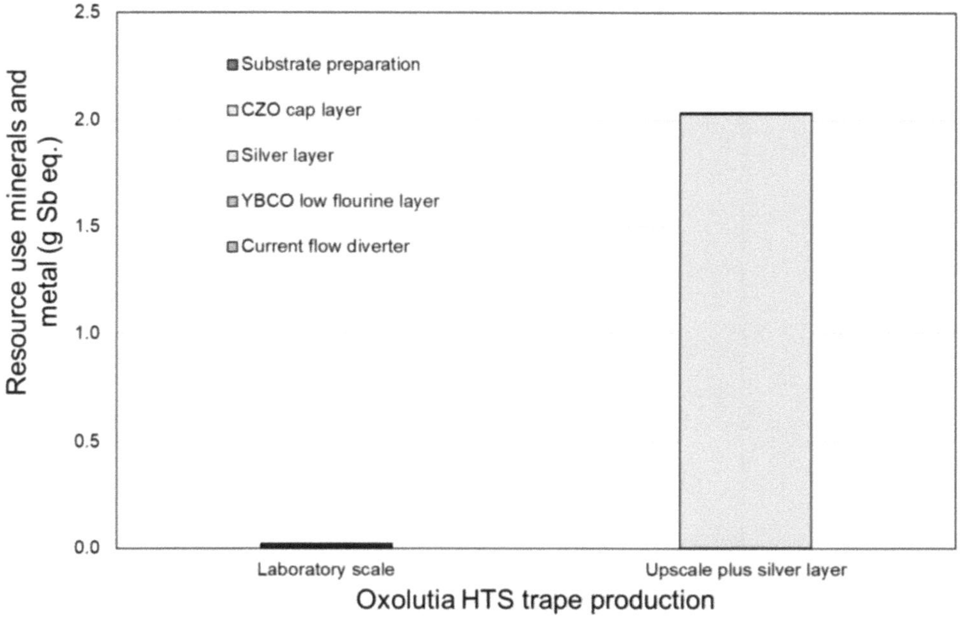

Figure 3.20: Impact comparison of the current yield at laboratory scale and the assumed upscaled yield at industrial scale with an additional silver layer in the impact category resource use minerals and metal.

Figure 3.21 shows the change in cumulative energy demand of Oxolutia tapes. Currently, this is around 133 kWh per metre of tape, which reflects the energy intensity of this type of production. This value is significantly higher than that of the current THEVA tape, which only requires 25.6 kWh per metre. Even considering an increased yield, the Oxolotia production process does not become less energy intensive than the THEVA process. Due to the added silver layer, the upscaled Oxolutia production process has a cumulative energy demand of 29.0 kWh. Compared to the next generation production of THEVA, which has a cumulative energy demand of 16.6 kWh, nearly twice the amount of energy would be consumed throughout the supply chain of the Oxolutia process.

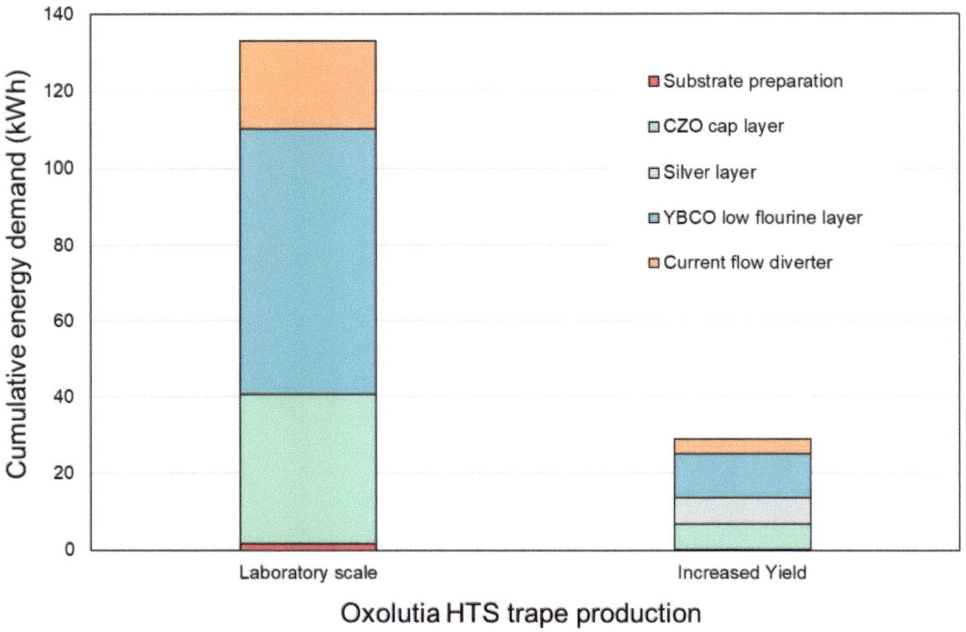

Figure 3.21: Impact comparison of the current yield at laboratory scale and the assumed upscaled yield at industrial scale with an additional silver layer in the impact category cumulative energydemand.

3.4.3 Tape Production Comparison

In the following, the two types of production will be compared across all impact categories. However, as the two tapes have different current carrying capacities, this must be taken into account when comparing them. The results of both tapes were therefore normalised to reflect the environmental impact per metre of tape per ampere of current carrying capacity. Figure 3.22 shows the direct comparison of both production processes as they currently are. The larger impact is shown as 100 % and the smaller impact is scaled accordingly.

It can be seen, that the current production of THEVA performs significantly better than the production of Oxolutia. On average, the environmental impact of the THEVA tape is just one fifth of that of the Oxolutia tape. Only in the categories eutrophication freshwater and resource use minerals and metals does the Oxolutia tape perform better. The difference in technology readiness levels is clearly noticeable here. The THEVA tape is already commercially available and the production process has already been optimised. Oxolutia's tape, on the other hand, is still in a development phase in which it is more a matter of proving the functionality of the production than optimising the production process.

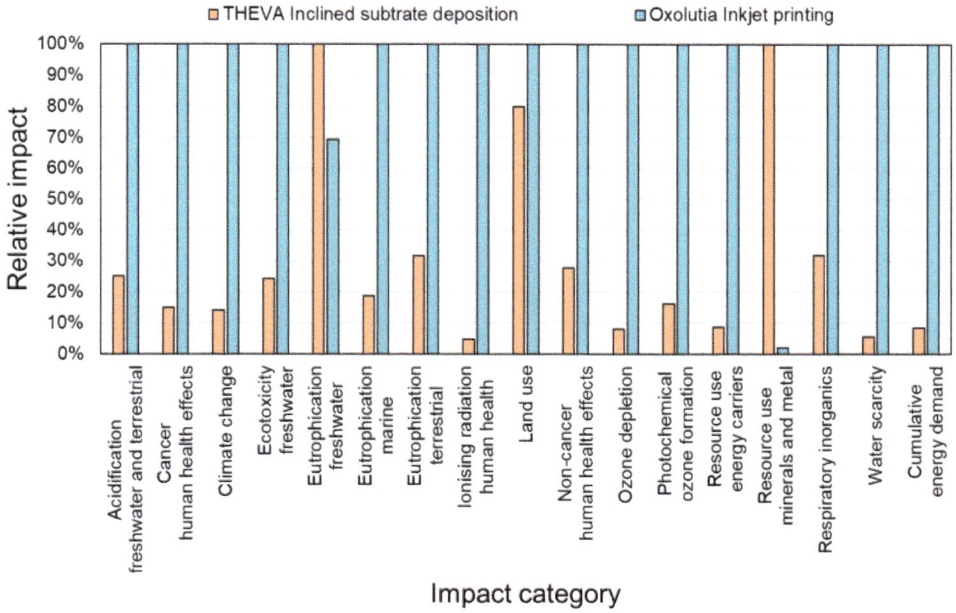

Figure 3.22: Relative comparison of the environmental impacts of the THEVA high-temperature superconductor tape production and the Oxolutia high-temperature superconductor tape production. The tape with the higher impact in each category is indicated by the given value of 100 %. The corresponding impact of the respective other tape is given as a relative share of 100 %.

In order to take into account possible future developments of both tapes, the results of the scenario analyses are also compared with each other. Figure 3.23 shows the comparison of THEVA tape with next generation production with Oxolutia tape with a higher yield and an additional silver layer. Here, the differences between the two tapes are smaller on average. However, contrary to the current production comparison, the next generation THEVA production does outperform the upscaled Oxolutia production process in every single environmental impact category.

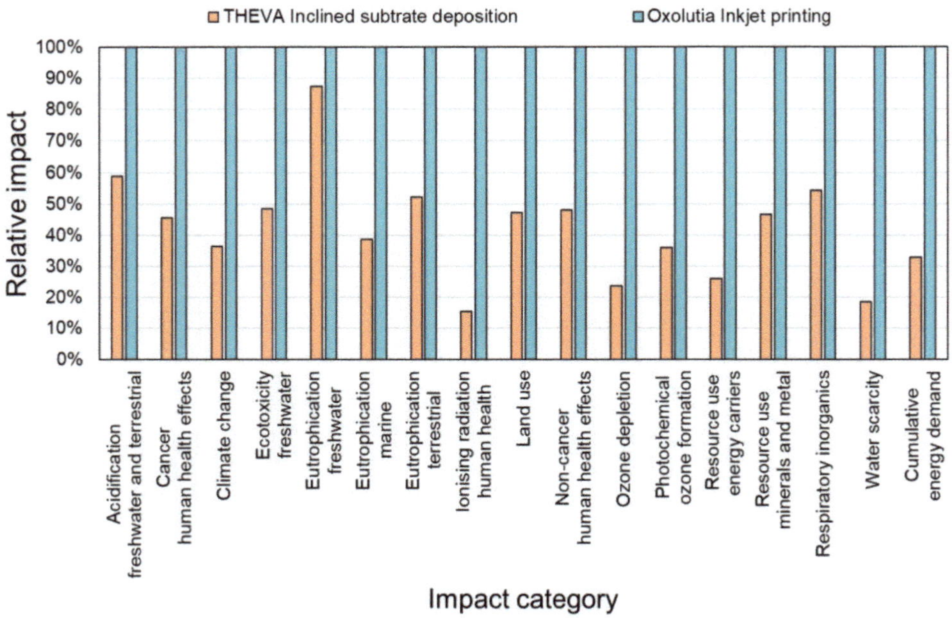

Figure 3.23: Relative comparison of the environmental impacts of the THEVA high-temperature superconductor tape next generation production and the Oxolutia high-temperature superconductor tape production with an upscaled yield with an additional silver layer. The tape with the higher impact in each category is indicated by the given value of 100 %. The corresponding impact of the respective other tape is given as a relative share of 100 %.

This is caused by the additional silver layer that has a particularly high impact in the categories ecotoxicity freshwater and resource use of minerals and metal. In these two categories, the current THEVA production performs worse than the Oxolutia tape for exactly this reason. Due to the added silver layer on the future Oxolutia tape, this difference between both tapes is not existing anymore resulting in higher environmental impacts of the Oxolutia tape.

3.4.4 Comparison with a Conventional Copper Conductor

In addition to comparing different high-temperature superconductor tapes, these tapes are compared with a conventional copper conductor in terms of their environmental impact. For this purpose, another product system was created for the production of a conventional copper conductor. Figure 3.24 shows the system boundaries of the copper conductor production considered in this comparison. The ecoinvent process of the global copper market was used for copper supply. This process obtains copper from both primary and secondary sources. The share of secondary copper amounts to 28.8 %. In addition, a wire drawing process is taken into account, which shapes the copper into a copper wire.

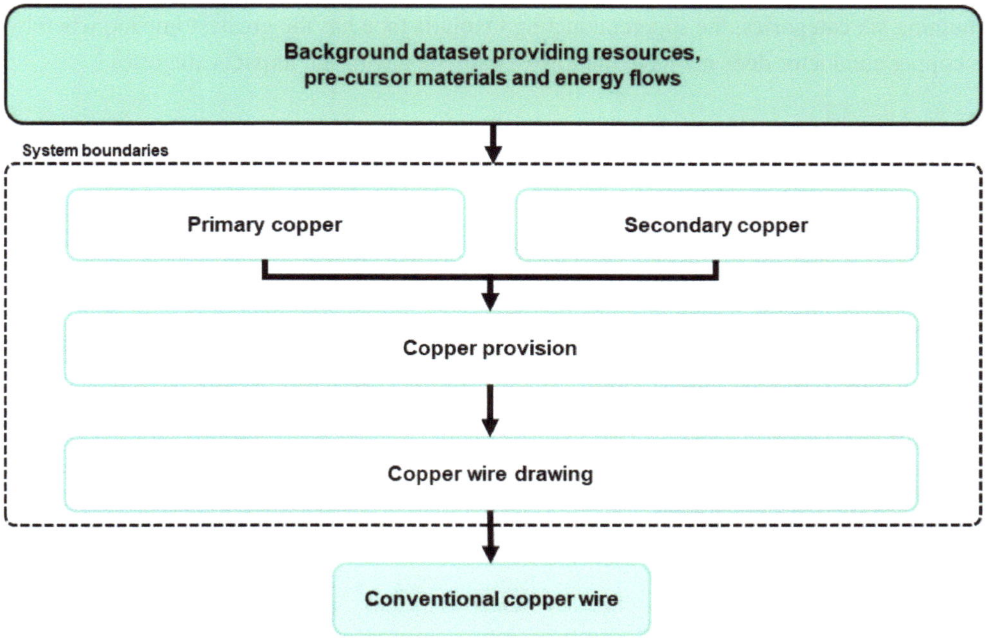

Figure 3.24: System boundaries of the production of a conventional copper conductor wire.

The conductivity of a copper wire depends on the diameter of the wire. For this comparison, a copper conductor was chosen which has the same current-carrying capacity as the high-temperature superconductor from THEVA. This corresponds to a current carrying capacity of 600 A. For a copper conductor to conduct this current with sufficiently low losses (> 20 W/m), it needs a diameter of about 2 cm. With a density of 8.92 g/cm³, this corresponds to a copper mass of about 2.8 kg per metre of copper wire.

For the comparison of high-temperature superconductors and copper conductors, it must be taken into account that the two superconductors already differ in terms of their current-carrying capacity. The environmental impacts of all three conductors in this comparison were therefore normalised by dividing the results of the life cycle impact assessment of each conductor by its respective current-carrying capacity. This makes it possible to find out which conductor production has the greatest environmental impact per metre per ampere of current-carrying capacity.

Figure 3.25 shows the relative results of the conductor comparison. In eleven of the total of 17 categories, the conventional copper conductor performs worst. In these eleven categories, the environmental impact of the Oxolutia superconductor is on average 62 % lower than that of the copper conductor. For the THEVA superconductor, the figure is as high as 85 %. In the remaining six categories, the superconducting Oxolutia tape has the greatest impact, whereby the copper conductor does not have the lowest value in any of the impact categories.

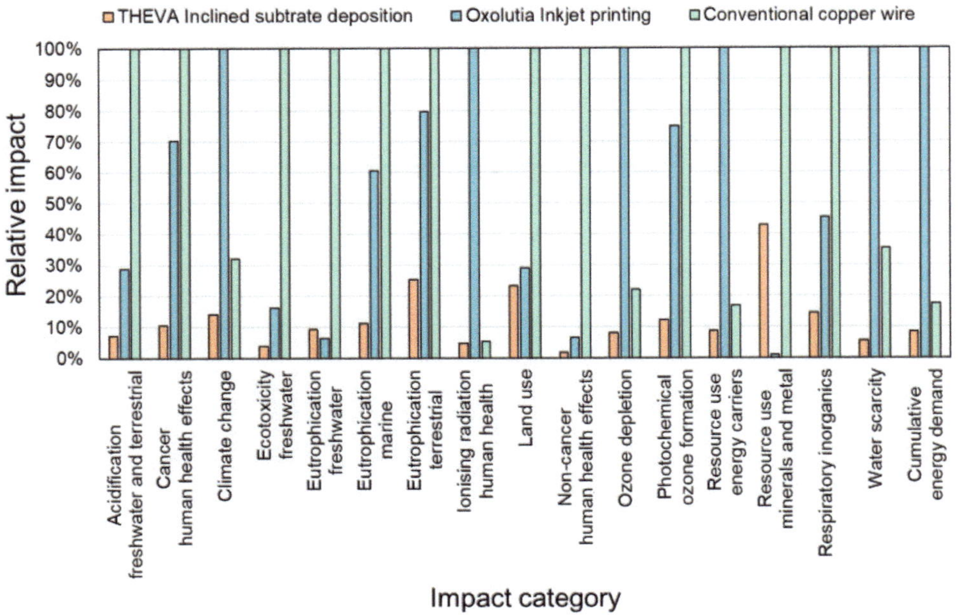

Figure 3.25: Relative comparison of the environmental impacts of the THEVA tape (orange), the Oxolutia tape (blue) and a conventional copper conductor wire (green). The conductor with the highest impact in each category is indicated by the given value of 100 %. The corresponding impact of the other conductors is given as a relative share of 100 %.

3.5 Interpretation

In this study, an LCA of two different production methods for manufacturing high-temperature superconductor tapes was carried out. The environmental im-pacts were investigated in a cra-dle-to-gate approach, so that the life cycle was only considered from the extraction of the raw materials to the finished pro-duction of the superconductors. The two production chains under consideration are at different technology readiness levels. While THEVA's tape is already commercially available, Oxolutia's production is still in a research and devel-opment stage. For both tapes, the data was provided directly by the manufac-turers and has a high data quality. In the case of THEVA, the production site was also visited several times to validate the data.

A contribution analysis was carried out for both tapes to find out which pro-duction steps have the greatest impact on the environment. Furthermore, an uncertainty analysis was carried out to increase the robustness of the results. A sensitivity analysis should lead to a better understanding of the overall system, while a scenario analysis should look at potential future changes in the produc-tion processes. The aim of the study was to be able to make recommendations on how the production of both companies can be made more environmentally friendly.

In the case of THEVA tape, the silver layer and the superconducting rare earth barium copper oxide layer were the main factors responsible for the environ-mental impact of the production chain, having an average share of 30 % and 39 %, respectively. At the same time, a sensitivity analysis showed that the overall system reacts most strongly to changes in these two layers. The prob-lem here is, that both layers cannot be reduced in thickness without affecting the quality of the tape. A thinner silver layer would result in less or even non-sufficient stabilisation of the tape. The superconducting layer itself does cor-relate quasi-linearly with the current-carrying capacity of the entire tape. A thinner superconducting layer would therefore reduce the current-carrying ca-pacity, which subsequently leads to more superconductor tapes being required in the use phase. This could result in a rebound effect in which the increased tape requirement could offset the savings from the layer reduction.

Another possibility for savings is to obtain a consistent layer thickness with less material and energy input. For this purpose, THEVA provided data regard-ing a planned future change in the production processes. These changes to the production process reduce greenhouse gas emissions by 1.9 kg and energy con-sumption by 9 kWh per meter of tape. A one-kilometer superconducting cable like the AmpaCity cable in Essen, Germany, requires about 150 km of 4 mm wide high-temperature superconductor tape. The proposed changes to the pro-duction process considered could therefore save a total of 95 metric tons of CO_2 eq. and 450 MWh of cumulative energy demand compared to the current production process. Such savings may simplify the use of superconducting technologies in the future by providing environmental benefits. For this pur-pose, however, it is necessary to consider the use phase of the superconductors. Compared to conventional conductors, they have the advantage of a higher current density and no ohmic losses, at least in DC applications, but they require constant cooling for this.

The further investigation of alternative ways to produce superconductors is an important component. The inkjet printing of the company Oxolutia examined in this study is still on a laboratory scale. Furthermore, it was identified that the environmental impact is mainly due to the energy consumption during the printing of the yttrium barium copper oxide layer. The use of renewable energy is therefore an obvious option to further reduce the environmental impact.

Nevertheless, the Oxolutia tape cannot compete with the THEVA tape and per-forms worse in most categories. Additionally, without a metal stabilisation layer the tape will not be protected against quench events. For a scenario anal-ysis, the yield was increased and an additional silver layer was added as a sta-bilisation to the tape architecture. While the increased yield significantly im-proves the environmental impacts, the additional silver layer causes the opposite effect and increases the environmental impacts. In total, however, the increased yield has a more significant influence and the environmental impacts could be reduced by about 70 % in most of the impact categories.

Furthermore, it must also be taken into account that the environmental impact of the sapphire substrate in this study is represented only by the provision of the required amount of Al_2O_3. No energy flows, material consumptions or losses are considered. All of these would further increase the environmental impact. Although literature values for the production of sapphire substrate were available, these were not used as they are very outdated inventory data. If these data had been used, however, the environmental impact of Oxolutia tape would be a factor of 3 to 40 higher, depending on the considered category.

THEVA tape also deals with relatively uncertain process data in substrate pro-duction. No literature values are available for the production of Hastelloy® C-276. Only the material content and two generic rolling processes from the Ecoinvent database were taken into account in order to map the environmental impacts. To make the results of future studies even more robust, one recom-mendation is therefore to collect more detailed inventory data for substrate production.

If future decision-makers of energy technologies are faced with the question which high-temperature superconductor tape they should use from an environ-mental point of view, the decision should be made primarily based on the en-vironmental categories, which have a high robustness. In this study, categories were identified that have a low uncertainty of outcome. Likewise, categories were also identified whose uncertainty are so high that they should not play a role for future decision-makers for the time being.

Overall, this LCA has shown that reductions in the environmental impact of high-temperature superconductor tape production cannot be achieved through changes to the architecture, but through increased material and energy effi-ciency. Manufacturers can now use this information to make their production more environmentally friendly.

One additional but crucial point that has to be considered is the recycling of superconducting tapes. Currently, the tapes are treated as scrap steel and are melted down. However, this can lead to a loss of the included materials. This point is especially important for the used materials that are considered critical by the European Commission such as the rare earths or magnesium. Losing these materials during the recycling process further increases their criticality. Therefore, tape manufacturers should make sure to design the tape in a way so that these materials can be recovered without any loss in quality.

Furthermore, it was shown that the production of high-temperature supercon-ductors is more environmentally friendly per ampere of current-carrying ca-pacity than that of a conventional copper conductor. With these results, how-ever, it must be taken into account that the function of conventional conductors and high-temperature superconductors is essentially the same, namely the transmission of electrical energy. However, the framework conditions of the two types of conductors are fundamentally different. A high-temperature su-perconductor only reaches its current-carrying capacity in a cooled state, while the conventional conductor has an ohmic resistance during its use. For this reason, a comparison of only the conductor productions is not meaningful and does not allow a statement on whether superconductors or conventional con-ductors are more environmentally friendly. Thus, the application of superconductors must also be taken into account in order to consider the entire life cycle. Only then, it can be assessed to what extent superconductors can substitute conventional conductors from an ecological point of view.

4 Life Cycle Assessment of a 10 kV High-Temperature Superconducting Cable System for Energy Distribution

Parts of this chapter have already been published in: A. Buchholz, M. Noe, D. Kottonau, E. Shabagin & M. Weil (2021): *Environmental Life-Cycle Assessment of a 10 kV High-Temperature Superconducting Cable System for Energy Distribution*, IEEE Transactions on Applied Superconductivity, Volume 31, Issue 5.

4.1 High-Temperature Superconducting Power Cables

In order to assess the environmental impact of high-temperature superconductors over the entire product life cycle, the application of superconductors in the use phase must be taken into account. Due to their electrical and magnetical properties, superconductors have a very broad field of application, so that the environmental impact in the use phase depends very much on the particular application. These potential applications include among others superconducting power cables, fault current limiters, high field magnets, fusion reactor coils, motors for electric aircraft, or generators for wind turbines [145] [146] [147] [148] [149] [150].

The focus of this work is on the application of superconductors in power cables. Firstly, the environmental impact can be compared to a conventional alternative. Secondly, there is already a large number of superconducting cables which are in test or even grid operation [151]. Table 4.1 shows a selection of superconducting cables from the last 15 years. Various studies have shown that superconducting cables can function reliably over a longer period of time and that they can even be economically advantageous over conventional cables [152] [153] [154] [155] [156] [157] [158].

Table 4.1: List with examples of superconducting cables around the world from the last 15 years (based on [159]).

Manufacturer	Place, Country	Year	Length	Specifications	HTS type
NKT*	Munich, Germany	N/A	12 km	110 kV, 500 MVA	YBCO
Nexans*	Chicago, USA	2021	N/A	12 kV	N/A
LS Cable	Seoul, S. Korea	2017	1000 m	22.9 kV	YBCO
Nexans	Essen, Germany	2014	1000 m	10 kV, 2.3 kA	BSCCO
Sumitomo	Yokohama, Japan	2013	240 m	66 kV, 1.8 kA	BSCCO
LS Cable	Icheon, S. Korea	2011	100 m	22.9 kV, 3.0 kA	BSCCO
LS Cable	Icheon, S. Korea	2009	500 m	22.9 kV, 1.3 kA	BSCCO
Nexans	Long Island, USA	2008	600 m	138 kV, 2.4 kA	BSCCO/YBCO
LS Cable	Gochang, S. Korea	2007	100 m	22.9 kV, 1.3 kA	BSCCO
Sumitomo	Albany, USA	2006	350 m	34.5 kV, 0.8 kA	BSCCO
Ultera	Columbus, USA	2006	200 m	13.2 kV, 3 kA	BSCCO
Sumitomo	Gochang, S. Korea	2006	100 m	22.9 kV, 1.3 kA	BSCCO

* = cable is in the process of planning and yet to be installed

Additionally, superconducting cables have several functional advantages over conventional cables. Superconducting cables have a higher current-carrying capacity and have low AC losses [157]. They are also smaller than conventional cables and can be installed more space-efficient, which is an advantage especially in inner-city areas [158].

One disadvantage compared to conventional cables is the constant need for cooling. This also requires a more complex cable design that allows a cooling medium to pass through the cable and not heat it up too much over the length of the cable. Figure 4.1 shows the general structure of three different cable types. The basic structure is similar for all cables. The innermost part is the former, through which the nitrogen is let in and pumped through the cable. The superconducting tapes and the electrical insulation are wound around the former. This layer is followed by a neutral conductor made of copper and finally the cryostat, which consists of two tubes. The liquid nitrogen is pumped back to the cooling unit through the inner tube, while there is a vacuum between the two cryostat tubes for thermal insulation. However, depending on the voltage level, the cable types differ in how many phases can be found in a cryostat. In a single-core cable, which is designed for applications in the range above 110 kV, there is only one phase inside a cryostat. The three-core cable is designed for applications from 30 kV to 110 kV and has all three phases in one cryostat, each on its own former. A three-phase, concentric cable can be used for the voltage range from 10 kV to 50 kV. Here, all three phases are also located in one cryostat, but they are all three wound around a single former, which allows for a particularly compact design.

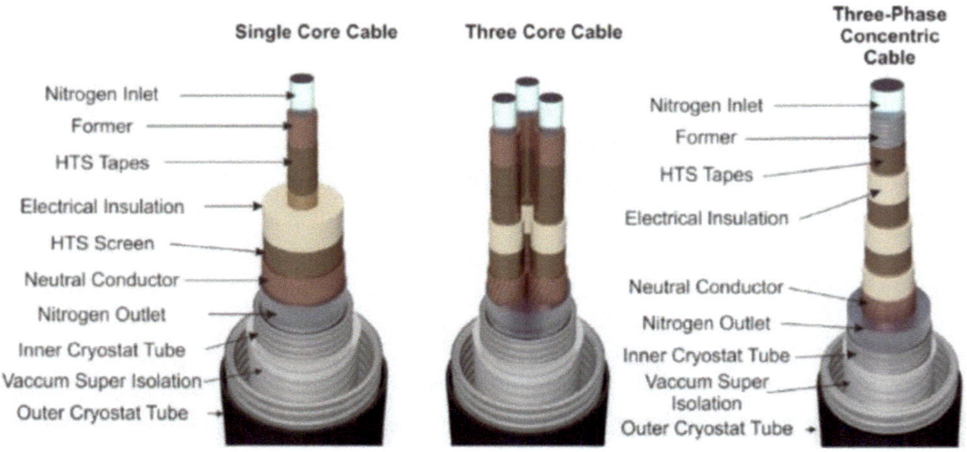

Figure 4.1: Three different types of superconducting cables and their various layers [159].

In this study, a three-phase concentric cable is considered. The superconducting medium-voltage cable from the AmpaCity project, which was installed in the city centre of Essen, Germany, and integrated into the distribution grid, is used as a case study. This project and the associated cable are presented in more detail below.

4.2 AmpaCity – a 1 km long, 10 kV, 40 MVA superconducting cable

The AmpaCity project started in 2011 with the participation of the cable manufacturer Nexans, the energy supply company RWE, and the Karlsruhe Institute of Technology. The project was funded by the Federal Ministry of Economics and Technology and the Project Management Jülich [160]. The industrial gas supplier Messer was responsible for the cooling system and the provision of liquid nitrogen [152].

Within the project, a superconducting three-phase concentric 10 kV cable was installed between two inner-city substations, replacing in future a conventional 110 kV cable. At the time of installation, the AmpaCity cable was the longest superconducting cable in the world with a length of one kilometre [160]. Furthermore, a superconducting fault current limiter was installed, making the AmpaCity cable the first superconducting cable in the world to be installed together with a fault current limiter [160].

The AmpaCity cable is a 10 kV three-phase concentric cable, which has a rated current of 2.31 kA and can transmit a rated power of 40 MVA. This superconducting cable could replace a 110 kV high-voltage cable with a rated current of 0.59 kA. Five conventional 10 kV cables with a rated current of 0.46 kA are used as redundancy for the AmpaCity cable. The AmpaCity cable went into operation in 2014. As the cable was operated without incident during the project phase, it has remained in the grid since then and is operated and maintained by the distribution grid operator Westnetz.

While there are already some economic studies on superconducting cables, environmental impacts are usually only rudimentarily considered in terms of the lack of thermal and electromagnetic impact as well as the lower space requirement [9] [161]. Due to the long operating time of the AmpaCity cable and the resulting long-term data regarding losses and cooling requirements, the cable is particularly well suited as a case study for the life cycle assessment of a superconducting cable.

4.3 Goal and Scope Definition

4.3.1 Goal

The aim of this study is a comparative life cycle assessment of a superconducting cable system similar to the AmpaCity cable. The superconducting cable is compared to conventional high and medium voltage copper cables. The comparison aims to analyse if and under which conditions superconducting cables can have ecological advantages over the conventional alternatives.

Using an attributional life cycle assessment approach, the superconducting cable is analysed in its status quo with the objective to identify the superconductor life cycle phase with the highest environmental impact. These findings will be used to give recommendations for future superconductor cables on how to improve the environmental performance.

As the AmpaCity cable was originally a test cable to show the feasibility of the use of superconducting cables in grids, there is still room for technological improvements. To consider potential future developments, an additional prospective life cycle assessment approach is used. In this prospective approach, potential future developments of the superconducting cable system are analysed. These potential developments cover changes in the transformer configuration of the cable system, the usage of a different cooling system, and the usage of a different electricity mix. These future developments are constructed in cooperation with experts from reseach and industry in order to minimise the uncertainty of the future scenario.

The comparison of the superconducting cable system with conventional alternatives requires precise data on both, the material input of all considered components and the losses during the use phase. The data should therefore be primary data from the manufacturers of the individual components or the operator of the cables. If this is not possible, secondary data from comparable components should be used.

As a superconducting cable system is a system of different components, data acquisition is a limiting factor of this study. The cable was produced by Nexans and is operated by Westnetz. The actual superconductors in turn came from a third source and are not from Nexans itself. The cooling system was designed by Messer who also provides the liquid nitrogen. There is no central data source and the individual stakeholders showed varying degrees of willingness to provide data for this study, so that the data quality for the individual components of the system fluctuates.

Furthermore, the superconductor in the actual AmpaCity cable is made of bismuth strontium calcium copper oxide. Since no data is available for this material, this study uses the gadolinium barium copper oxide superconductors from THEVA described in chapter 3.1 to approximate the environmental impact of the used superconductor. However, as this type of superconductor is considered the preferred type for energy applications anyway, the change of superconductor material is not expected to have a negative impact on the quality of the result, but rather is even a more accurate representation of future cable systems [89] [90].

The impact assessment methods Environmental Footprint 3.0 and Cumulative Energy Demand are chosen in this study, to calculate the environmental impacts of the cable systems. As a background database, ecoinvent 3.5 with the cut-off system is used. [86].

4.3.2 Scope

The product system under consideration includes all components necessary to transmit a power of up to 40 MVA between two substations in the inner-city area over a distance of one kilometre. It is taken into account that the voltage must be transformed from the maximum voltage of 380 kV to the medium voltage of 10 kV. This means that in addition to the cable under consideration, two transformers are also required in each case. The first transformer for the transformation from 380 kV to 110 kV and the second for the transformation from 110 kV to 10 kV. However, since these two transformers have the same power and are needed regardless of which cable system is selected, their impacts can be neglected for the actual comparison of the cable systems.

Figure 4.2 shows the system boundaries of the cable comparison, which shows the various options for transmitting the 40 MVA power between two substations. In option a), the electricity is transported with the help of a conventional 110 kV cable and transformed to 10 kV at substation B. In the case of option b), both voltage transformations down to 10 kV already take place at the first transformer station. The electricity is then transmitted to substation B by means of five 10 kV cables. The last option is the use of a superconducting cable. As with the conventional 10 kV cable, the voltage is transformed down to 10 kV at the first substation and then transmitted to the second substation. In this system, however, the necessary cooling system must also be taken into account. Additionally, a superconducting fault current limiter (SFCL) is installed to protect the superconducting cable from fault currents and thus must also be considered.

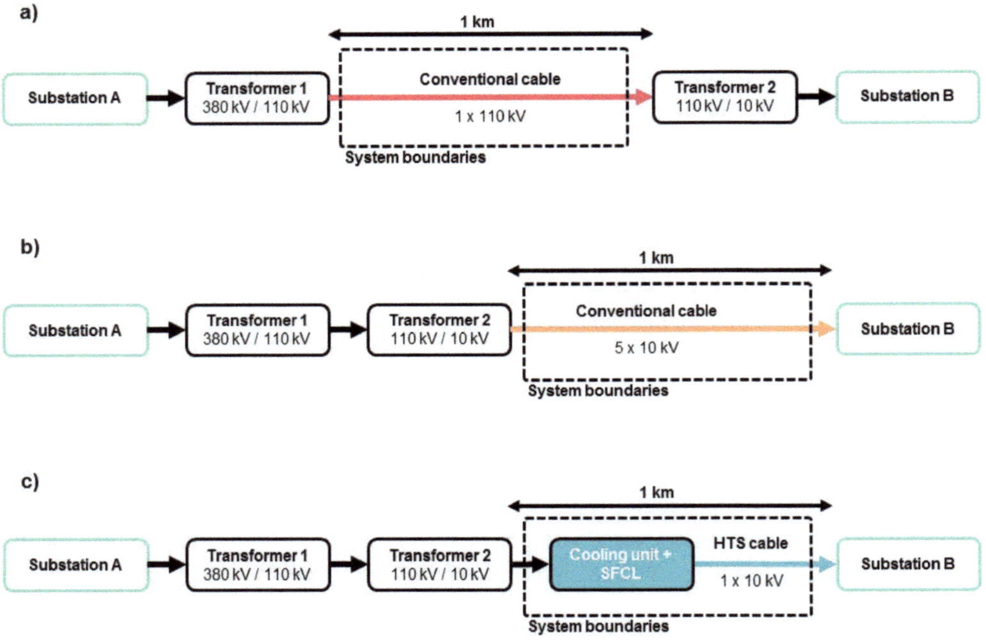

Figure 4.2: System boundaries of a) a 110 kV conventional cable, b) five 10 kV conventional cables and c) a high-temperature superconducting (HTS) cable, the fault current limiter, and its required cooling system.

The functional unit for all three cables is the annual transmitted electricity. Each cable is able to transmit a power of up to 40 MVA. The comparison is carried out using the cradle-to-grave approach. This means that in addition to the production, the use phase of all cables is considered. Where possible, the waste treatment of the individual materials is also taken into account to also include the end-of-life phase. In this context, all components are assumed to have a lifetime of 40 years.

To analyse under which circumstances a superconducting cable can be a viable alternative to conventional cables two additional comparisons are made. In the first additional comparison, the transformer configuration is changed so that the superconducting cable system uses only one transformer, while the conventional high voltage cable uses two transformers. The second additional comparison analyses the use of an alternative closed cooling system. In the case of the AmpaCity cable, an open cooling system is used. The term open means that the cooling energy is provided by evaporating liquid nitrogen. As the liquid nitrogen leaves this open system, it must be refilled regularly [152]. In a closed cooling system, liquid nitrogen is recooled electrically and does not leave the system.

Furthermore, the uncertainty of the data is analysed and taken into account via a Monte Carlo simulation.

4.4 Life Cycle Inventory Analysis

4.4.1 High-Temperature Superconducting Cable System

The superconducting cable system consists of a superconducting 10 kV cable, a fault current limiter and a cooling system. A separate life cycle inventory is created for each component of the system. In addition, the losses during the use phase must be taken into account, as these losses cause heat which is absorbed by the liquid nitrogen. The heated liquid nitrogen must then be recooled resulting in the consumption of liquid nitrogen. Therefore, the losses directly determine the amount of liquid nitrogen that must be provided during the use phase of the superconducting cable system. Figure 4.3 shows the detailed system boundaries for the superconducting cable system with all relevant unit processes that are necessary to ensure the transmission of a power of up to 40 MVA.

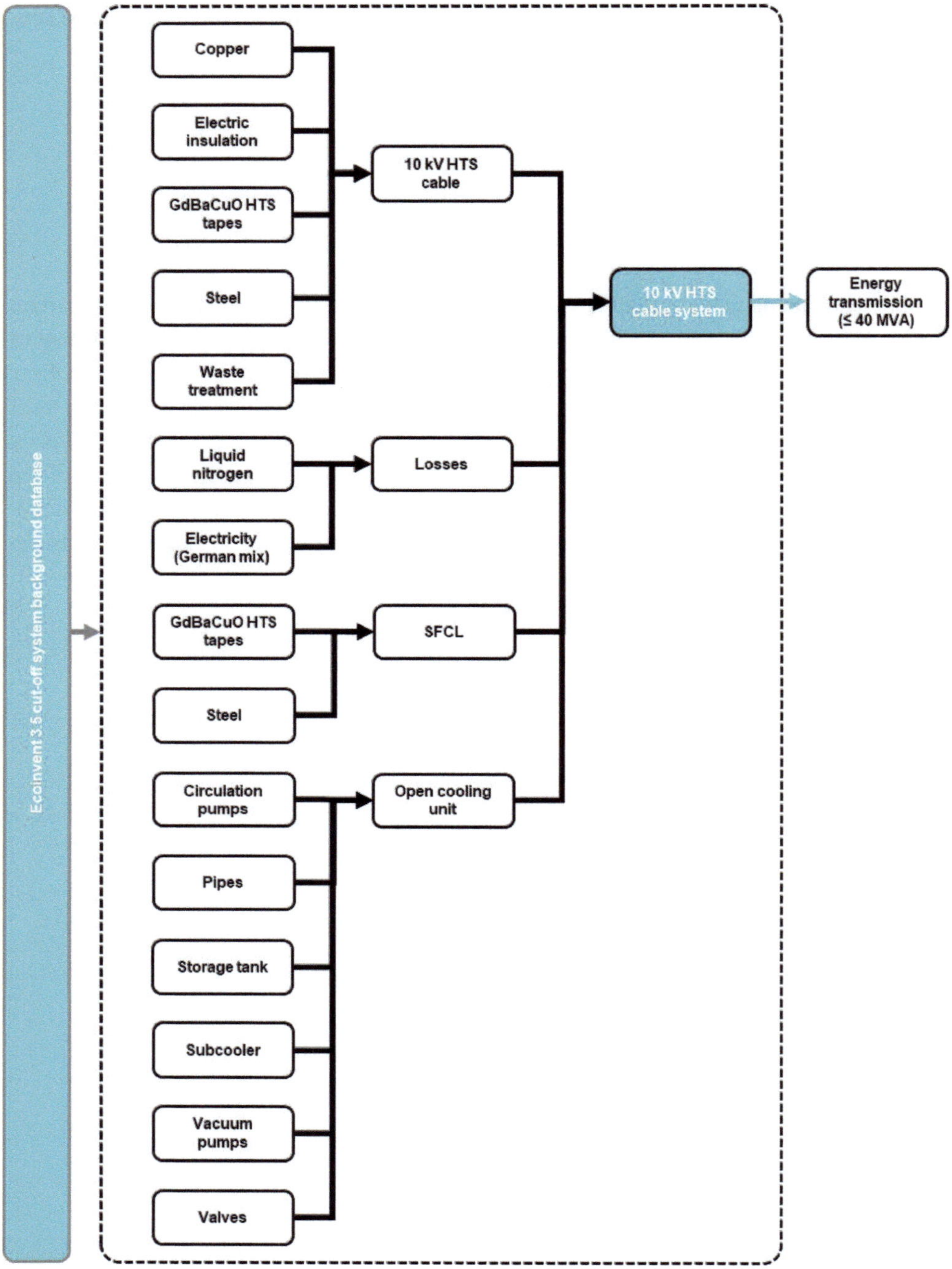

Figure 4.3: System boundaries of the 10 kV high-temperature superconducting cable system with an open cooling unit and a superconducting fault current limiter.

4.4.1.1 Losses and Liquid Nitrogen Demand

In addition to the materials for components, the losses that occur when transmitting the power must be calculated. These losses cause a heat input into the liquid nitrogen. Liquid nitrogen is then flowing in the subcooler to provide the cooling energy to recool the liquid nitrogen in the circulation with a heat exchanger. However, to enable continuous cooling, the nitrogen is constantly reproduced and replenished. Thus, the losses are directly related to the nitrogen consumption of the cable system. In total, three sources of loss occur in the cable system:

- Voltage-dependent losses P_{Volt}
- Current-dependent losses $P_{Current}$
- Thermal losses $P_{Thermal}$

The voltage-dependent losses are calculated from the angular frequency of the AC voltage ω, the capacity of the cable C, the nominal voltage U_N, and the loss factor tan δ. This type of losses is current-independent, constantly at 0.1 kW and can be calculated via equation 1. It is important to note, that this type of loss does not only generate a heat impact into the liquid nitrogen but must also be considered as grid losses.

$$P_{Volt} = \omega * C * {U_N}^2 * \tan\delta \qquad \text{(Eq. 1)}$$

The current-dependent losses come from several sources. In addition to the superconducting cable itself, the superconducting fault current limiter and the power supplies also have current-dependent losses. All current-dependent losses are also length-dependent. Since the cable is one kilometre long, in the following equation the factor 1 will not be included for reasons of clarity.

Current-dependent losses in the cable ($P_{Current,C}$) only occur in alternating current applications, due to alternating magnetic fields causing hysteresis losses in the superconductor and eddy current losses in the normal conductor [162]. These losses can be roughly estimated as follows via equation 2. Here, $P_{Current,C,Ic}$ represents the current-dependent losses at the maximum continuous current I_c of 2.31 kA.

$$P_{Current,C} = P_{Current,C,I_c} * \left(\frac{I}{I_c}\right)^3 \qquad \text{(Eq. 2)}$$

For reasons of simplicity, the current-dependent losses in the fault current limiter were estimated with the same behaviour using eqation 3. Since the amount of superconductor in the fault current limiter is small in comparsion to the cable this part should not have a large influence at all. In equation 3, $P_{Current,F,Ic}$ denotes the current-dependent losses at the maximum continuous current I_c of 2.31 kA.

$$P_{Current,F} = P_{Current,F,I_C} * \left(\frac{I}{I_C}\right)^3 \quad \text{(Eq. 3)}$$

The losses of the current leads can be divided into two sources. About 60% of the total losses are thermal losses, whereas the current-dependent losses account for about 40 % of the total losses of the current lead systems. For the calculation of the total losses of the current leads P_L, typical loss values of a current lead are assumed to be 45 W/kA [162].This value is multiplied by a rated current I_r of 1.8 kA, resulting in the maximum loss values at the rated current of a single current lead, of which there are six in the overall system. The total current-dependent losses of the current lead $P_{Current,L}$ result from equation 4. In this equation, $P_{L,Ir}$ denotes the typical losses of current leads at the rated current I_r.

$$P_{Current,L} = 0.4 * P_{L,I_r} * \left(\frac{I}{I_r}\right)^3 \quad \text{(Eq. 4)}$$

These three current-dependent loss sources can now simply be added up according to equation 5 to calculate the total current-dependent losses. As with the voltage-dependent losses, the current-dependent losses generate heat but they are also grid losses.

$$P_{Current} = P_{Current,C} + P_{Current,F} + P_{Current,L} \quad \text{(Eq. 5)}$$

In contrast to the voltage- and current-dependent losses, which cause the conductor itself to heat up, the thermal losses are an external heat input into the liquid nitrogen. These inputs occur at the cable cryostat, the cable terminations and the current leads.

As said before, the thermal losses of the current leads $P_{Thermal,L}$ account for about 60 % of the total current lead losses at the rated current I_r. They are calculated with equation 6.

$$P_{Thermal,L} = 0.6 * P_{L,I_r} \quad \text{(Eq. 6)}$$

The thermal losses of the cable cryostat $P_{Thermal,Cryo}$ are about 1.3 W/m. For the thermal losses of the cable terminations $P_{Thermal,T}$, 100 W are assumed. The total thermal losses result from the sum of the partial losses, as can be seen from equation 7.

$$P_{Thermal} = P_{Thermal,L} + P_{Thermal,Cryo} + P_{Thermal,T} \quad \text{(Eq. 7)}$$

In addition to the heat input by the cable losses, there is a further heat input into the overall system through the circulation pumps. According to information from Messer, this additional heat input through the pumps P_{Qpump} is about 1.6 kW.

From these losses, it is now possible to calculate the total heat input P_{Qtotal} that must be cooled off by the liquid nitrogen using equation 8.

$$P_{Qtotal} = P_{Volt} + P_{Current} + P_{Thermal} + P_{Qpump} \quad \text{(Eq. 8)}$$

The total losses are provided in Table 4.2.

Table 4.2: Hourly heat input by loss type for different cable loads as well as the resulting total generated heat. All values are given in kW [163].

Loss type	10% S_n	30% S_n	50% S_n	70% S_n	90% S_n	S_n = 40 MVA
Heat generated by current-dependent losses $P_{Current}$	0.00	0.07	0.35	0.95	2.01	2.76
Heat generated by voltage-dependent losses P_{Volt}	0.10	0.10	0.10	0.10	0.10	0.10
Thermal losses $P_{Thermal}$	1.69	1.69	1.69	1.69	1.69	1.69
Pump heat impact P_{Qp}	1.60	1.60	1.60	1.60	1.60	1.60
Total heat input P_{Qtotal}	3.39	3.47	3.74	4.34	5.40	6.15

The value P_{Qtotal} can now be used to calculate the amount of liquid nitrogen needed to cool the heat input. The enthalpy of vaporisation of liquid nitrogen is 198.6 kJ/kg at 77.3 K. This means that a heat input of 198.6 kW evaporates a nitrogen mass of 1 kg in one second. Hence, 1 kW of losses over a period of one hour evaporates a nitrogen mass of about 18.1 kg.

This mass of liquid nitrogen can now be converted into an energy equivalent by calculating the energy consumption of the liquid nitrogen production. The energy demand to produce one kg of liquid nitrogen is about 0.56 kWh. This energy demand can be multiplied with the mass of liquid nitrogen that is evaporated by P_{Qtotal} resulting in the energy consumption of the liquid nitrogen P_{LN2} that is caused by the losses of the cable system. Table 4.3 shows the values for the heat input P_{Qtotal} for different cable loads as well as the resulting liquid nitrogen demand and liquid nitrogen energy consumption P_{LN2}.

Table 4.3: Hourly heat input, liquid nitrogen demand and energy consumption during liquid nitrogen production for different cable loads.

	10% S_n	30% S_n	50% S_n	70% S_n	90% S_n	S_n = 40 MVA
Total heat input P_{Qtotal} (kW)	3.39	3.47	3.74	4.34	5.40	6.15
Liquid nitrogen demand (kg)	61.44	62.74	67.63	78.53	97.82	111.36
Liquid nitrogen production energy P_{LN2} (kW)	34.31	35.13	37.88	43.98	54.78	62.36

The energy consumption during liquid nitrogen production can now be included into the system losses in order to quantify the electric losses of the cooling unit. However, there are other loss sources that must be considered to calculate the total losses of the superconducting system P_{total}. Firstly, the current-dependent losses $P_{Current}$ and the voltage-dependent losses P_{Volt} also cause grid losses. Thus, they must be taken into account. Additionally, the electricity consumption of the circulation pumps $P_{pump,c}$ and the vacuum pumps $P_{pump,v}$ must be considered. These are 4 kW per hour and 5 kW per hour, respectively. Thus, the total system losses P_{total} can be calculated with equation 9. Table 4.4 shows detailed values for each of these losses as well as the total system losses for various cable loads.

$$P_{total} = P_{LN2} + P_{Current} + P_{Volt} + P_{pump,c} + P_{pump,v} \qquad \text{(Eq. 9)}$$

Table 4.4: Hourly electricity consumption by source for different cable loads as well as the resulting total system losses P_{total}. All values are given in kW.

	10% S_n	30% S_n	50% S_n	70% S_n	90% S_n	S_n = 40 MVA
Liquid nitrogen production energy P_{LN2}	34.31	35.13	37.88	43.98	54.78	62.36
Current-dependent losses $P_{Current}$	0.00	0.07	0.35	0.95	2.01	2.76
Voltage-dependent losses P_{Volt}	0.10	0.10	0.10	0.10	0.10	0.10
Circulation pump energy consumption $P_{pump,c}$	4.00	4.00	4.00	4.00	4.00	4.00
Vacum pump energy consumption $P_{pump,v}$	5.00	5.00	5.00	5.00	5.00	5.00
Total HTS system losses P_{total}	43.73	44.54	47.57	54.31	66.25	74.63

The total system losses are the source of environmental impacts during the use phase and thus essential for the life cycle assessment of a superconducting cable system. Using these loss values, the hourly environmental impacts can be calculated. However, the functional unit of this study is the annual transmitted power of the cable systems and therefore, the annual losses must be considered by calculating the losses for every hour of the year. Over the course of one year however, a cable does not have a constant but rather a fluctuating load. To take this into account, the annual load factor m_a is introduced [164] [165].

This load factor m_a describes the ratio between the transmitted load and the peak load multiplied by time and is calculated from equation 10. In this equation, t_a describes the hour of the year ($t_a \leq 8760$), S is the transmitted load ($S \leq 40$ MVA) and S_N is the peak load ($S_N = 40$ MVA).

$$m_a = \frac{\int_0^{t_a} S(t) * dt}{S_N * t_a} \qquad \text{(Eq. 10)}$$

Using the load factor, annual load profiles can be calculated via equation 11. These load profiles provide descendingly sorted load values for each hour of the year while also considering load fluctuations. Figure 4.4 shows normalised annual load profiles for different load factors.

$$y(x) = 1 - (1 - {m_a}^2) * x^{m_a} \qquad \text{(Eq. 11)}$$

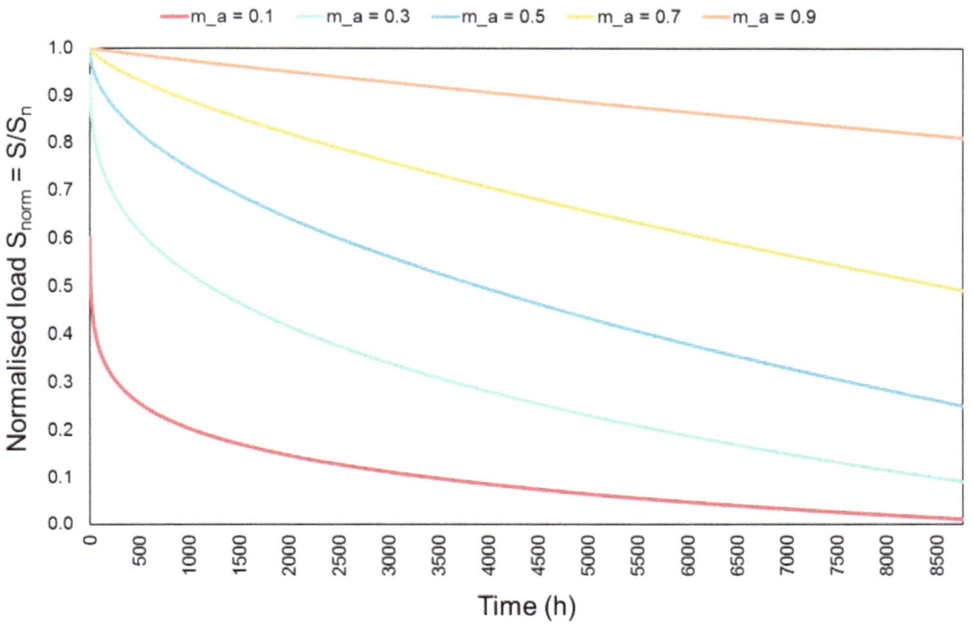

Figure 4.4: Normalised annual load profiles for five different load factors.

As these load profiles provide a load value for each hour of the year, these load values can be used to calculate the system loss energy for each hour of the year using equations 1-9 resulting in a loss profile.

Figure 4.5 shows the loss profile of the superconducting cable system for a load factor of $m_a = 0.7$ over the period of one year. It is also shown how the individual loss sources contribute to the total system losses. While the pumps consume electricity at site, the energy consumption due to the pump heat impact and the thermal losses happens at the air separation unit that produces the liquid nitrogen.

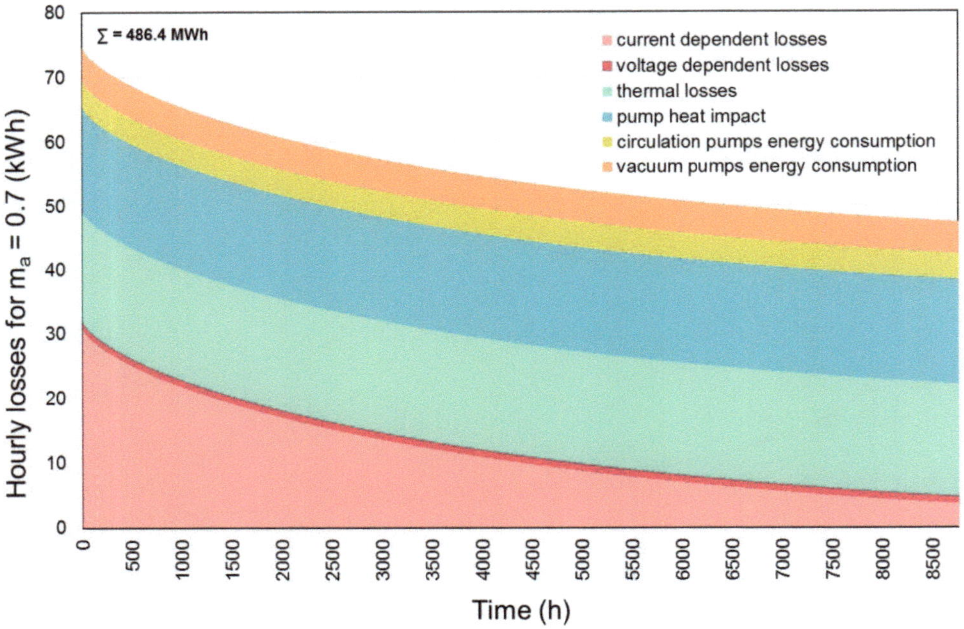

Figure 4.5: Hourly losses of the superconducting cable system by source over the course of one year using a load factor of $m_a = 0.7$.

If the integral of this loss profile is formed, the annual total system losses can be calculated. With a load factor of $m_a = 0.7$, this annual energy loss is 486.40 MWh. The same calculation can be performed for each possible load factor ($0 \leq m_a \leq 1$) to determine the annual loss energy as a function of the load factor. Figure 4.6 shows the contribution of each loss type to the total system losses for every load factor. Eeach bar in this figure represents the sum of 8760 hourly values based on the loss profile of the corresponding load factor.

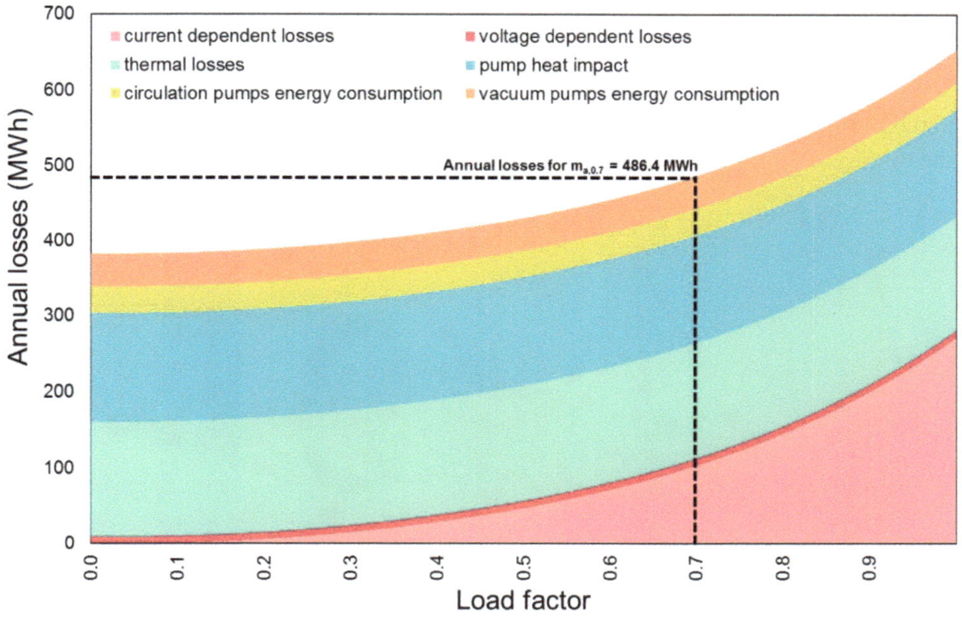

Figure 4.6: Annual loss energy of the superconducting cable system by loss source as a function of load factor. Dashed lines indicate the annual losses at $m_a = 0.7$ and thus corresponds with the integral of the loss profile shown in Figure 4.5.

4.4.1.2 Three-Phase Concentric Cable

In addition to the losses during the use phase, the materials of the individual components must also be taken into account. No primary data could be acquired from Nexans for the cable. Therefore, the material requirements had to be calculated using the dimensions of a model and literature values. Figure 4.7 shows the schematic three-phase concentric structure of the cable. The former consists of a corrugated steel tube. Since these are standardised according to DIN EN 10220, standard values are used for the dimensions of the former [166]. A tube with an outer diameter of 26.9 mm is selected. The weight of such a tube is 326 kg per km. The superconducting gadolinium barium copper oxide tapes and the dielectric consisting of polypropylene laminated paper are wound alternately around the former. For the three phases, 41 km, 52 km, and 61 km of four-millimetre-wide superconducting tapes are required. The polypropylene laminated paper consists of approximately 42% kraft paper and 58 % polypropylene by mass. The three layers of dielectric each have a total thickness of 3.7 mm to 4.2 mm. The mass of both components can be calculated using the density of polypropylene and kraft paper, respectively. In total, about 1.2 t of kraft paper and 1.6 t of polypropylene are necessary to produce the dielectric for one kilometre of cable. The neutral conductor layer is made of copper and has a thickness of about 2.8 mm. With a density of 8.96 g/cm³, this corresponds to a mass of about 7.5 t of copper per kilometre of cable. As with the former, a standardised corrugated steel sheet diameter according to DIN EN 10220 is assumed for the cryostat. An outer diameter of 76.1 mm is selected for the inner cryostat and an outer diameter of 114.3 mm for the outer cryostat, as these diameters correspond most closely to the dimensions of the cable model. In total, this results in a mass of 5.8 t and 14.5 t of steel per kilometre of cable for the two cryostat tubes.

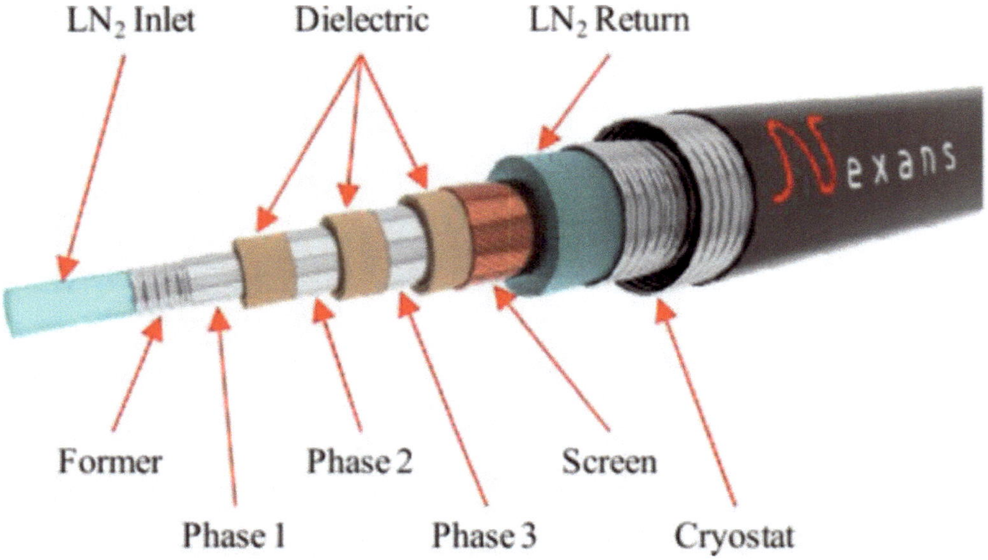

Figure 4.7: Schematic of the three-phase concentric cable design of the AmpaCity cable [160].

It is important to mention that no information regarding the production of the cable by Nexans has been disclosed. Therefore, only the pure material requirement of the cable and the energy input for the production of the superconductor tapes, the drawing of copper wires and the rolling of steel are considered. The actual production as well as the installation of the cable can not be considered in this study. An overview of the entire life cycle inventory for the superconducting cable can be found in Table 4.5.

Table 4.5: Life cycle inventory of a one kilometre long 10 kV three-phase concentric superconducting cable.

Material	Input Amount	Unit
Copper	7.6	t
GdBaCuO superconducting tape	154.0	km
Kraft paper	1.2	t
Polypropylene	1.6	t
Steel	23.2	t
Waste treatment copper	-7.6	t
Waste treatment steel	-23.2	t
Material	**Output Amount**	**Unit**
10 kV three-phase concentric cable	1	km

4.4.1.3 Superconducting Fault Current Limiter

The superconducting cable system as considered in this study also contains a superconducting fault current limiter. As the cable itself, the fault current limiter was produced by Nexans and thus, no data regarding the production was disclosed. Therefore, only the required material to provide the current limiting function is considered in this study. For this purpose, literature values that describe the material demand of a superconducting fault current limiter for a three-phase concentric superconducting cable are taken as a reference [146].

According to these literature values, the fault current limiter consists of 21 modules, each with ten twelve-millimetre-wide superconductor tapes, each with a length of 19 metres. Thus, a total of about 4 km of tapes is required. In addition, a shunt resistor made of steel is connected in parallel. This resistor also consists of 21 modules of 12 mm wide and 35 mm thick steel tapes. The total mass of steel required for the fault current limiter is about 100 kg. In addition to the material demand, the rolling of the steel into tapes and the necessary disposal of the steel are taken into account. The entire inventory is shown in Table 4.6.

Table 4.6: Life cycle inventory of a superconducting fault current limiter for a superconducting fault current limiter.

Material	Input Amount	Unit
GdBaCuO superconducting tape	4.0	km
Steel	102.9	kg
Steel, sheet rolling	102.9	kg
Waste treatment steel	-102.9	kg
Material	**Output Amount**	**Unit**
Superconducting fault current limiter	1	item

4.4.1.4 Open Cooling System Components

The superconducting cable system uses an open cooling system designed by Messer [152]. Figure 4.8 shows a schematic of this cooling system. Here, liquid nitrogen is evaporated in a subcooler under vacuum conditions to generate the cooling energy.

The liquid nitrogen in the circuit is then cooled via a heat exchanger in the subcooler and pumped through the cable cryostat and back into the subcooler via circulation pumps. Evaporation of the nitrogen results in a continuous loss of liquid nitrogen in the system. This nitrogen must be replenished at regular intervals. A storage tank with a volume of 50 m³ is available for this purpose. This storage tank provides the liquid nitrogen for the circulation and is refilled every two to three weeks.

For reasons of redundancy, two circulation pumps and three vacuum pumps are required for the cooling system [152]. The entire cooling system therefore consists of a storage tank, a subcooler with heat exchanger, two circulation pumps, three vacuum pumps and the necessary pipes as shown in Table 4.7. The material requirements for the individual components are taken from an already existing study [167]. The life cycle inventories of the individual components are given in appendix D. A lifetime of 40 years is assumed for the entire cooling system, as for all other components of the cable system.

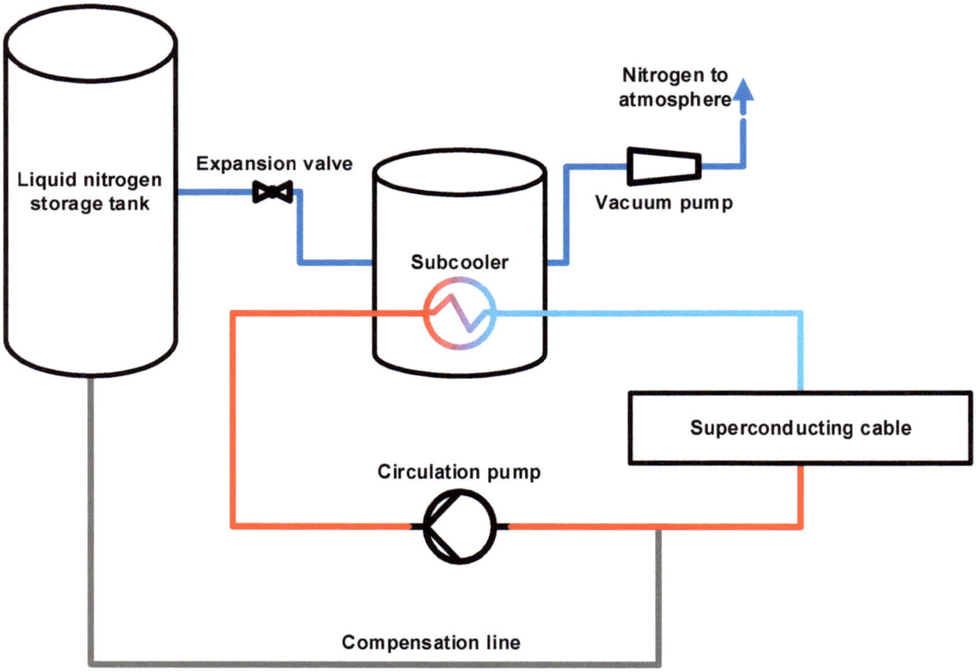

Figure 4.8: Schematic representation of the AmpaCity open cooling system [152].

Table 4.7: Life cycle inventory of an open cooling system [167].

Material	Input Amount	Unit
Circulation pumps	2	#
Liquid nitrogen storage tank	1	#
Piping	1	#
Subcooler	1	#
Vacuum pumps	3	#
Material	**Output Amount**	**Unit**
Open cooling system	1	#

4.4.2 Conventional Cable Systems

The superconducting cable system is compared to two different conventional cable systems. Firstly, a conventional 110 kV cable and secondly, a medium voltage system consisting of five 10 kV cables.

Figure 4.9 shows the system boundaries for a conventional 110 kV cable with all relevant unit processes (see Figure 4.2). This system consists of the conventional cable itself, all its precursor components and the ohmic losses of the cable.

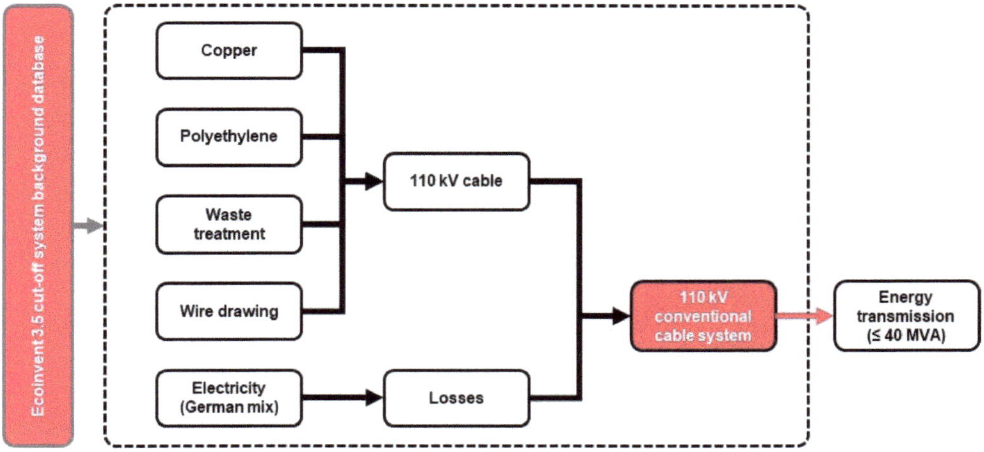

Figure 4.9: System boundaries of the 110 kV conventional cable system.

Figure 4.10 shows the system boundaries for the conventional medium voltage alternative consisting of five 10 kV cables. The relevant unit processes cover the cables and their components as well as their losses during the use phase.

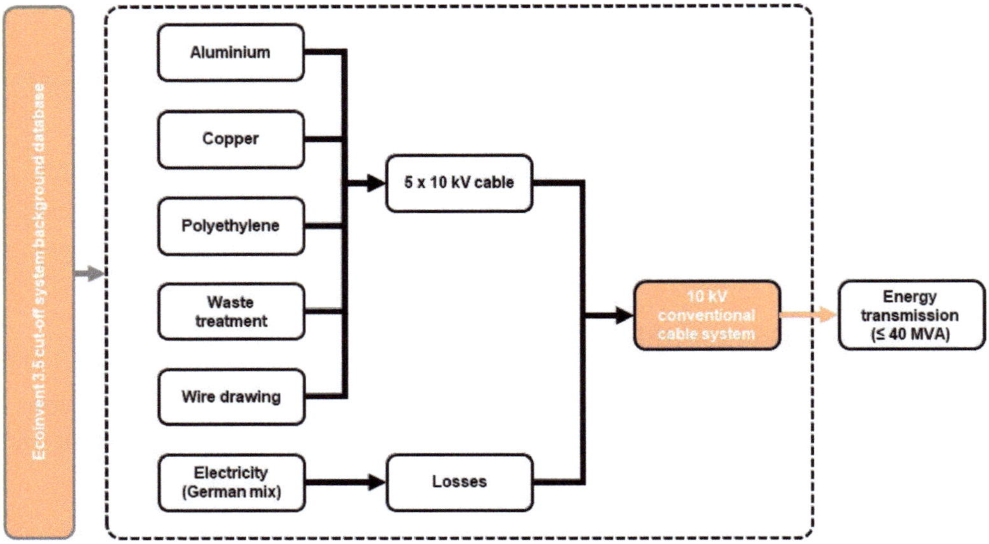

Figure 4.10: System boundaries of the 10 kV conventional cable system.

4.4.2.1 Losses

As with the superconducting cable, the losses of the conventional cable systems must be considered for the environmental impact of the use phase. For both conventional cable systems, losses occur due to the ohmic resistance R of the conductor material. Furthermore, these losses P_{conv} are dependent on the transmitted current I and the resistance R of the cable. For the 110 kV cable, a resistance of 95.5 mΩ/km is assumed, while this value is 60.1 mΩ/km for the 10 kV cable [162]. Using equation 12, the losses P_{conv} can be calculated for both cable systems.

$$P_{conv} = 3 * I^2 * R \tag{Eq. 12}$$

In case of the conventional medium voltage cable system, equation 12 must be multiplied by five as there are five cables in the system.

As there are no other losses to be considered for the conventional cable systems, the resulting values equal the total system losses of the conventional cables. These system losses can now be compared to the superconducting cable system losses as shown in Table 4.8.

Table 4.8: Hourly losses of the two conventional cable systems for different cable loads. For comparison, the total system losses of the superconducting system are also shown. All values are given in kW.

	10% S_n	30% S_n	50% S_n	70% S_n	90% S_n	S_n = 40 MVA
Losses of the conventional 110 kV cable $P_{conv, 110}$	0.13	1.14	3.16	6.19	10.23	12.63
Losses of the five conventional 10 kV cables $P_{conv, 10}$	1.92	17.32	48.10	94.29	155.86	192.42
Total HTS system losses P_{total}	43.31	44.13	46.88	52.98	63.78	71.36

As with the superconducting cable system, the loss profile of the conventional cables can be calculated using the load factor m_a and the load profiles. Figure 4.11 shows the loss profiles of both conventional cable systems in comparison with the superconducting cable system loss profile. Due to the constant cooling of the superconducting cable system, there are losses even when there is no cable load. Thus, at low loads the superconducting cable has annual losses of more than about 400 MWh while the conventional cable systems have losses of less than 200 MWh for $m_a \leq 0.3$. However, the losses of the conventional medium voltage system increase significantly faster with an increasing load than the losses of the superconducting cable. Above a load factor of $m_a = 0.47$, the superconducting cable system has less yearly loss energy than the conventional medium voltage cable system, making it the preferable option in terms of energy efficiency. Compared to a conventional high voltage cable, the superconducting cable cannot compete regardless of the load factor.

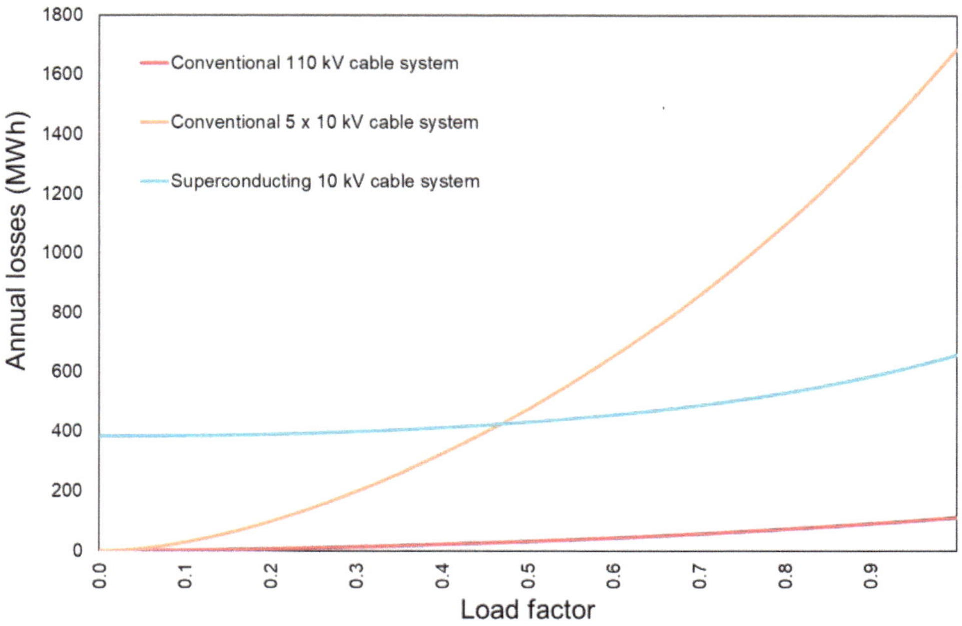

Figure 4.11: Annual loss energy as a function of the load factor for each of the three cable systems.

As with the superconducting cable, the losses of the conventional cable system are the cause of environmental impacts during the use phase. However, to analyse the impacts of conventional cable systems over the entire life cycle the production phase and thus the material requirements must also be considered.

4.4.2.2 Conventional Cable System Components

For the conventional 110 kV cable, a cable of the type N2XS(FL)2Y RM/50 1x300 mm² is selected. This cable consists of a copper conductor with polyethylene insulation. According to the cable manufacturer NKT, this cable requires a copper mass of 6.4 kg/m [168]. As the cable length in this comparison is one kilometre, this results in a total mass of 6.4 t of copper. For the insulation layer, a mass of about 2 t polyethylene is assumed for a layer thickness of 18 mm. This value is based on the density of polyethylene of about 0.9 g/cm³. Table 4.9 shows the entire life cycle inventory for a conventional 110 kV cable including wire drawing and waste treatment processes.

Table 4.9: Life cycle inventory of a conventional 110 kV cable [169].

Material	Input Amount	Unit
Copper	6.4	t
Copper wire drawing	6.4	t
Polyethylene	2.0	t
Waste treatment copper	-6.4	t
Material	**Output Amount**	**Unit**
Conventional 110 kV cable	1	km

The 10 kV cable is type NA2XS2Y RM/35 1x630 mm². This cable uses both aluminium and copper and also has polyethylene insulation [170]. The amount of aluminium required is about 1.8 t, while a copper mass of about 394 kg is necessary. For the insulation, a total of about 279 kg of polyethylene is used. The entire life cycle inventory for one conventional 10 kV cable is given in Table 4.10.

Table 4.10: Life cycle inventory of a conventional 10 kV cable [169].

Material	Input Amount	Unit
Aluminium	1.8	t
Copper	394	kg
Copper wire drawing	2.2	t
Polyethylene	279	kg
Waste treatment aluminium	-1.8	t
Waste treatment copper	-394	kg
Material	**Output Amount**	**Unit**
Conventional 10 kV cable	1	km

4.5 Life Cycle Impact Assessment

In accordance with the system boundaries described in chapter 4.3, the environmental impacts are calculated and presented below. As a basis, the life cycle inventories described in chapter 4.4 are used to calculate the impact of the material demand of each cable system. The use phase is analysed based on the losses of each cable system. For the superconducting cable, these losses translate into a liquid nitrogen demand. Thus, the environmental impacts of the use phase of the superconducting cable system depend on the amount of liquid nitrogen that needs to be produced. Therefore, it is important to know the environmental impacts of the production of liquid nitrogen.

Additionally, the losses must be considered as grid losses. This also causes environmental impacts for the superconducting cable system as the cable is the cause for the electricity being left unused.

For the conventional cable systems, the use phase impacts occur only due to the grid losses. However, the losses occur on different voltage levels for both considered conventional cable systems. This difference is important as the environmental impacts of 1 kWh high voltage electricity differ slightly from the impacts of 1 kWh of medium voltage electricity in the ecoinvent database. This is because the medium voltage electricity process in the database does consider transformation losses.

However, as there is data uncertainty for the electricity mixes as well as for the liquid nitrogen production, the environmental impacts of 1 kWh electricity and 1 kg of liquid nitrogen are also uncertain. Therefore, a Monte Carlo simulation is conducted for the production of 1 kg liquid nitrogen and for 1 kWh of high or medium voltage electricity from the German electricity mix in the database.

This results in a range for the impact factors for the use phase of each of the cable systems rather than one specific value which allows to consider data uncertainty in the comparison. Table 4.11 shows the impact factors for 1 kg of liquid nitrogen, 1 kWh of medium voltage electricity and 1 kWh of high voltage electricity. For each parameter, the base result as well as the minimum and maximum result of the Monte Carlo simulation are shown.

Table 4.11: Impact factors of the processes relevant for the environmental impacts of the use phase for the individual cable systems. Shown is a range of values calculated by a Monte Carlo simulation, which takes into account the uncertainty of the underlying data.

		1 kg liquid nitrogen			1 kWh electricity medium voltage			1 kWh electricity high voltage		
Impact category	**Unit**	**Min**	**Base**	**Max**	**Min**	**Base**	**Max**	**Min**	**Base**	**Max**
Acidification terrestrial and freshwater	mol H+ eq.	1.8E-03	2.6E-03	3.9E-03	2.7E-03	4.5E-03	7.3E-03	3.6E-03	4.6E-03	6.5E-03
Cancer human health effects	CTUh	1.5E-09	2.1E-09	8.2E-09	3.0E-09	3.7E-09	1.3E-08	2.2E-09	3.2E-09	1.8E-09
Climate change	kg CO_2eq.	2.8E-01	3.6E-01	5.2E-01	4.8E-01	6.4E-01	8.4E-01	6.5E-01	6.5E-01	6.8E-01
Ecotoxicity freshwater	CTUe	5.8E-02	6.7E-02	2.1E-01	1.2E-01	1.2E-01	5.6E-01	1.1E-01	1.1E-01	1.4E+00
Eutrophication freshwater	kg Peq.	1.2E-04	5.2E-04	2.3E-03	3.1E-04	9.2E-04	5.1E-03	3.6E-04	9.3E-04	4.1E-03
Eutrophication marine	kg Neq.	2.3E-04	3.1E-04	4.8E-04	4.4E-04	5.6E-04	7.7E-04	4.9E-04	5.6E-04	6.7E-04
Eutrophication terrestrial	mol Neq.	6.2E-03	9.6E-03	1.5E-02	9.5E-03	1.7E-02	2.8E-02	1.1E-02	1.7E-02	2.6E-02
Ionising radiation	kBq U-235eq.	1.7E-02	7.6E-02	8.6E-01	3.9E-02	1.3E-01	2.2E+00	3.8E-02	1.4E-01	1.1E+00
Land use	Pt.	1.8E-01	2.3E-01	4.5E-01	3.1E-01	4.1E-01	8.8E-01	3.5E-01	4.1E-01	7.0E-01
Non-cancer human health effects	CTUh	2.2E-08	2.7E-08	1.0E-07	3.4E-08	4.8E-08	1.4E-07	3.8E-08	4.8E-08	1.8E-07
Ozone depletion	kg CFC-11eq.	9.2E-09	1.3E-08	2.5E-08	1.7E-08	2.3E-08	4.9E-08	1.9E-08	2.3E-08	4.3E-08
Photochemical ozone formation	kg NMVOCeq.	3.6E-04	4.6E-04	7.7E-04	6.4E-04	8.2E-04	1.2E-03	7.4E-04	8.3E-04	1.1E-03
Resource use, energy carriers	MJ	3.8E+00	5.0E+00	7.3E+00	6.6E+00	8.9E+00	1.2E+01	8.6E+00	9.0E+00	9.6E+00
Resource use, mineral and metals	kg Sbeq.	1.1E-07	1.5E-07	9.4E-07	1.8E-07	2.1E-07	1.5E-06	1.1E-07	1.3E-07	8.0E-07
Respiratory inorganics	Disease incidences.	9.5E-09	1.5E-08	2.3E-08	1.5E-08	2.6E-08	4.4E-08	1.8E-08	2.6E-08	4.0E-08
Water scarcity	m^3 deprived	4.4E+01	5.7E+01	8.3E+01	7.4E+01	9.9E+01	1.3E+02	9.8E+01	1.0E+02	1.1E+02
Cumulative energy demand	kWh	1.1E+00	1.6E+00	4.8E+00	2.0E+00	2.9E+00	3.6E+00	2.8E+00	2.9E+00	3.3E+00

4.5.1 Superconducting Medium Voltage Cable vs. Conventional High Voltage and Medium Voltage Cables

The annual loss energy for each load factor can be calculated via the load profiles. Based on the loss energy, the required amount of liquid nitrogen can be calculated as described in equation 9 in chapter 4.4. By multiplying the required amount of liquid nitrogen with the corresponding impact factor from Table 4.11, the environmental impact of the cooling of the superconducting cable system are calculated. The same can be done for the grid losses that are multiplied with the corresponding impact factor of the medium voltage electricity to calculate the entire impacts of the use phase. In addition to these use phase impacts, the environmental impacts of the material demand result in the total environmental impacts of the superconducting cable system. However, since a lifetime of 40 years is assumed for each component only 1/40 of the total material impacts must be considered with regard to the functional unit of annual electricity transmission.

4.5.1.1 Contribution Analysis for the 10 kV High-Temperature Superconducting Cable System

The environmental impacts of the superconducting 10 kV cable system can be divided into use phase impacts and material demand impacts. The use phase impacts are based on the liquid nitrogen demand and the grid losses. The material demand considers the materials of the cable itself, the superconducting fault current limiter, and the cooling unit.

Figure 4.12 shows the contribution analysis for the impact category climate change as a function of the load factor. In total, the annual greenhouse gas emissions are between 260.8 t CO_2 eq. and 435.6 t CO_2 eq. per year. The contribution of the use phase to these impacts is between 94.6 % for low loads and 96.7 % for full load.

In terms of materials, the superconducting cable itself has the greatest impact. While it has a share of 4.3 % of the total impact at low load factors, this share drops to 2.6 % as the load factor increases. The material requirements of the fault current limiter and the cooling unit thus account for less than 1 % of the total impact each.

For the environmental impacts of the use phase, the current-dependent losses and the voltage-dependent losses must be considered twice. Firstly, they cause environmental impacts due to their heat impact into the liquid nitrogen. This results in an evaporation and consequently a reproduction of liquid nitrogen. Secondly, they cause environmental impacts due to being grid losses that do not reach the end user. Therefore, electricity was generated which caused environmental impacts. However, since the cable losses are the cause of this electricity never being used the environmental burden is also allocated to the cable.

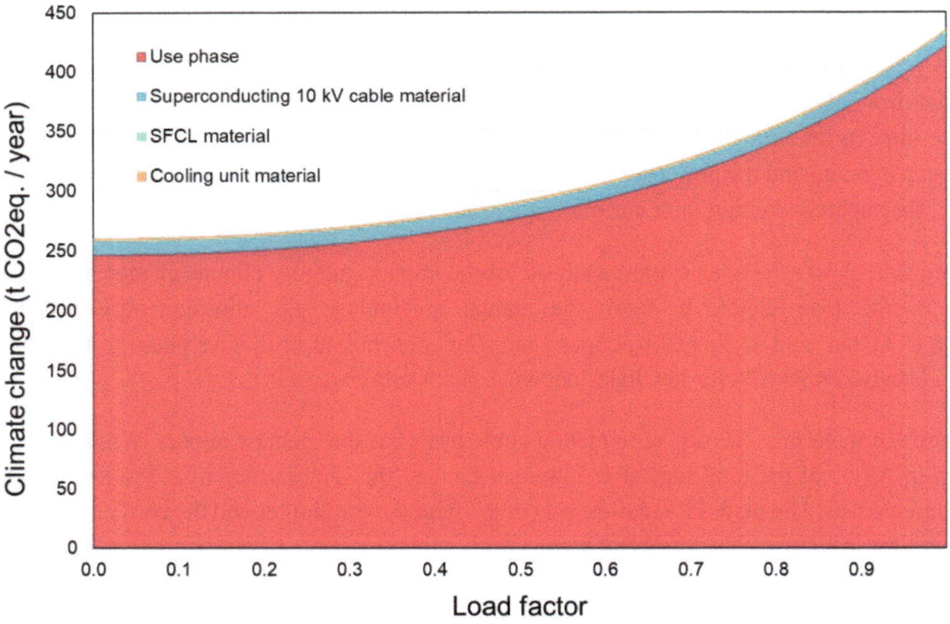

Figure 4.12: Climate change impacts of the superconducting 10 kV cable system by source as a function of load factor.

Figure 4.13 shows the contribution analysis results for each impact category for the load factor $m_a = 0.7$. This load factor is chosen as it represents a load that is commonly used in calculations of power supply companies. For this load factor, the use phase has a mean share of 79.5 % across all impact categories. In twelve out of 17 categories, the use phase has a share of more than 80 %. Additionally, the use phase has the highest share in all impact categories except for resource use, mineral and metals. This category is dominated by the superconducting cable itself with a share of 87.9 %.

On average, the materials for the cable have a share of 16.5 % across all impact categories. However, in eleven of the 17 categories the share is less than 10 %. The material demand for the fault current limiter is less than 1.0 % on average, while the cooling unit material accounts for a mean share of 3.0 %. Only in the categories cancer human health effects and ecotoxicity freshwater, the cooling unit reaches a share of moren than 10 %.

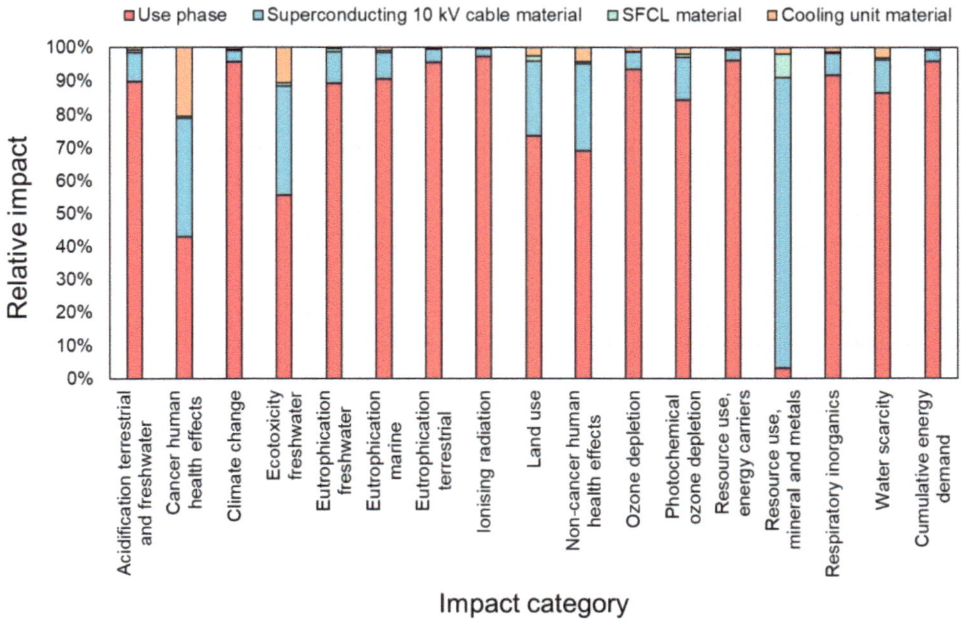

Figure 4.13: Relative impact of the use phase and the material demand of the different system components of the superconducting cable system for each impact category at a load factor of $m_a = 0.7$.

The impacts of the cooling unit materials can mostly be traced back to the liquid nitrogen storage tank which has a mean share of 77.7 % in the impacts of the cooling unit materials as can be seen in Figure 4.14. The vacuum pumps have the second highest average share with a value of 8.4 %. However, the subcooler has a mean impact share of 7.8 % and is thus only slightly less impactful than the vacuum pumps. The circulation pumps cause 5.1 % of the material impacts on average. Piping and miscellaneous components only have an average share of less than 1.0 %.

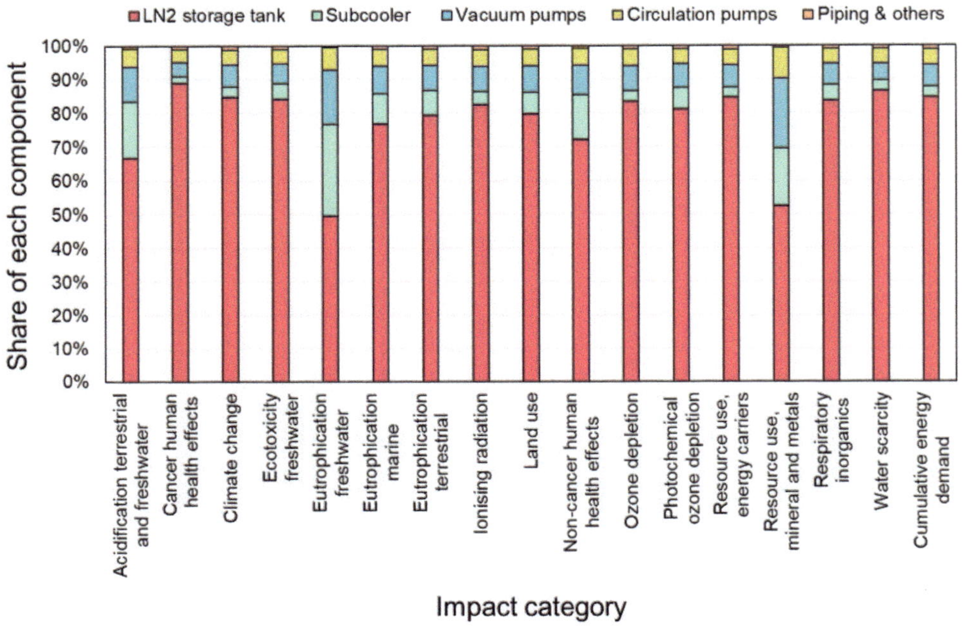

Figure 4.14: Share of each component of the cooling system of the superconducting cable system to the total impact of the material demand for a load factor of $m_a = 0.7$.

4.5.1.2 Contribution Analysis for the 110 kV Conventional Cable System

The conventional 110 kV cable has hardly any losses at low load and thus hardly any environmental impacts due to the use phase, as shown in Figure 4.15. The ohmic losses of the cable are responsible for less than 1 % of the total effects at very low loads. However, at a load factor of $m_a = 0.1$ the share of the use phase is 58.3 % already. At maximum load, the losses cause a total of 98.8 % of the impacts in the climate change category. This drastic increase is due to the quadratic relationship between load and losses. In total, the greenhouse gas emissions increase with an increasing load from 0.87 to a maximum of 73.1 t CO2 eq. per year.

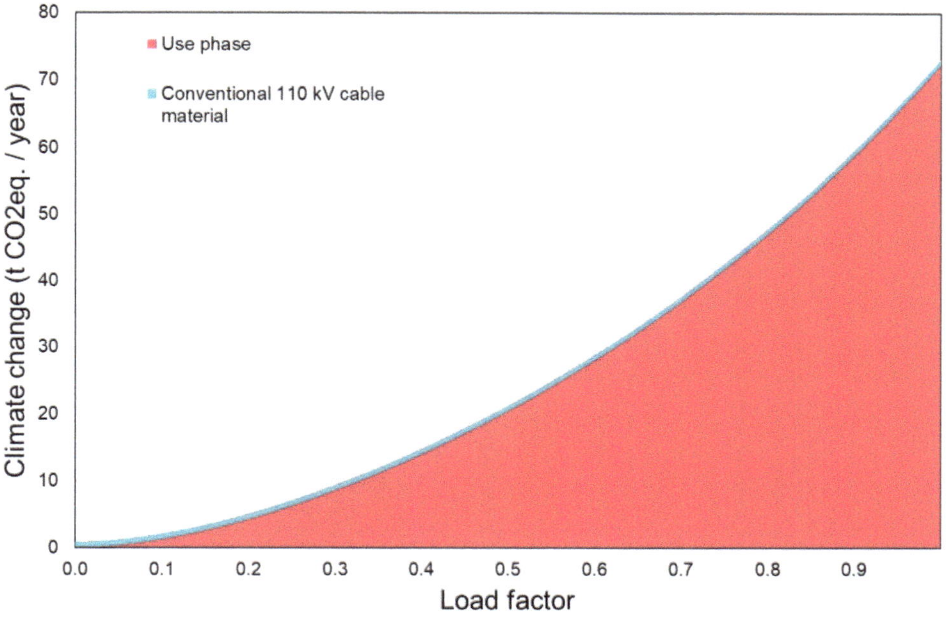

Figure 4.15: Climate change impacts of the conventional 110 kV cable by source as a function of load factor.

At the power supply company load $m_a = 0.7$, the average share of the use phase of the conventional 110 kV cable is 77.5 % across all impact categories as shown in Figure 4.16. Thus, the material demand for the cable accounts for 22.5 %. However, in the categories ecotoxicity freshwater, non-cancer human health effects, and resource use of mineral and metals, the share of the material demand is above 50 %. For the latter, the material demand causes 97.7 % of the impact and thus almost the entire impact in the resource use category.

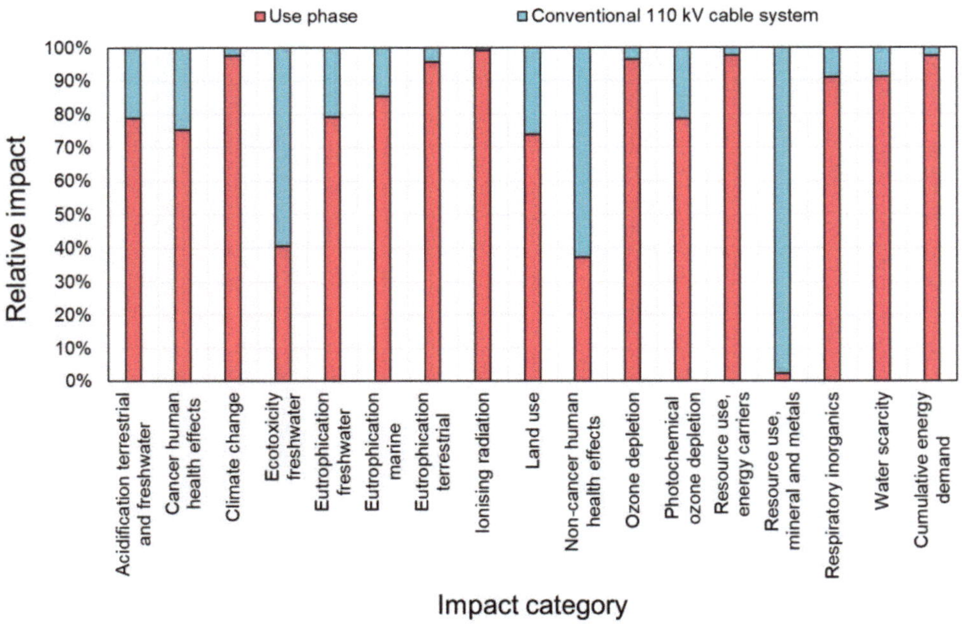

Figure 4.16: Relative impact of the use phase and the material demand of the conventional 110 kV cable system for each impact category at a load factor of $m_a = 0.7$.

4.5.1.3 Contribution Analysis for the 5 x 10 kV Conventional Cable System

As with the conventional high voltage cable, the climate change category of the medium-voltage cable system with five 10 kV cables is dominated by the losses of the cables. Figure 4.17 shows the load-dependent effects for this medium-voltage cable system.

At very low load factors, the impacts of the use phase are negligible as the losses of the cables are also low. However, already from a load factor of $m_a = 0.2$, the losses are responsible for over 95 % of the total greenhouse gas emissions per year. At full load, the losses cause 99.7 % of the climate change impact of the conventional 5 x 10 kV cable system.

Looking at the absolute values, the conventional medium voltage cable system has the highest climate change impacts of all three cable systems. At full load, the greenhouse gas emissions are about 1089.2 t CO_2 eq. per year and thus significantly higher than the impacts of the conventional 110 kV cable system (73.1 t CO_2 eq. per year) and the superconducting 10 kV cable system (435.6 t CO_2 eq. per year).

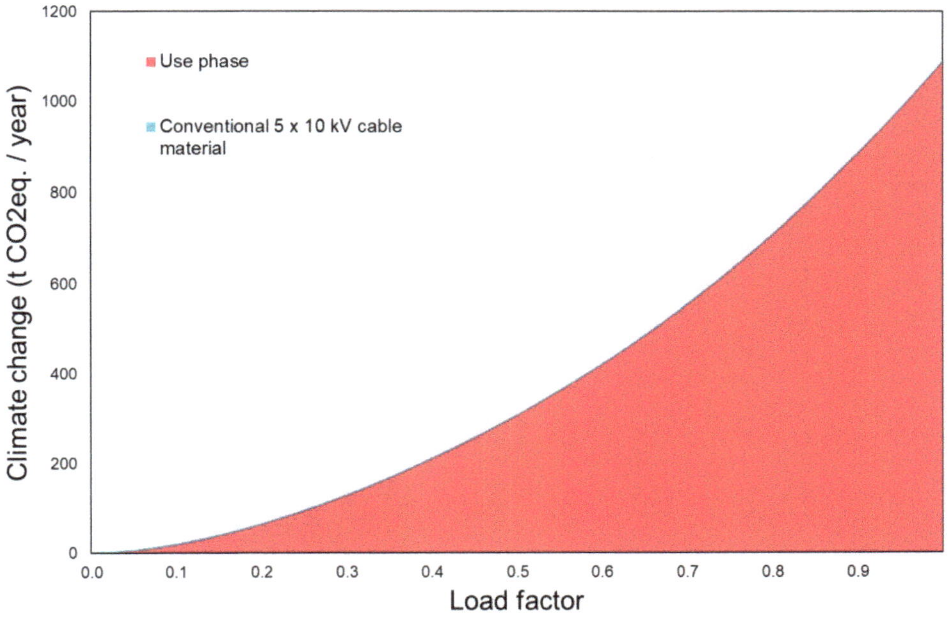

Figure 4.17: Climate change impacts of the conventional 5 x 10 kV cable by source as a function of load factor.

Due to the high losses of the conventional 5 x 10 kV cable system during the use phase, the average share of the use phase at $m_a = 0.7$ is 96.1 % across all impact categories as shown in Figure 4.18. The material demand of the cables has a share of less than 5.5 % in all impact categories except for the resource use of mineral and metals. In this category, the material demand accounts for 39.4 % of the total impacts.

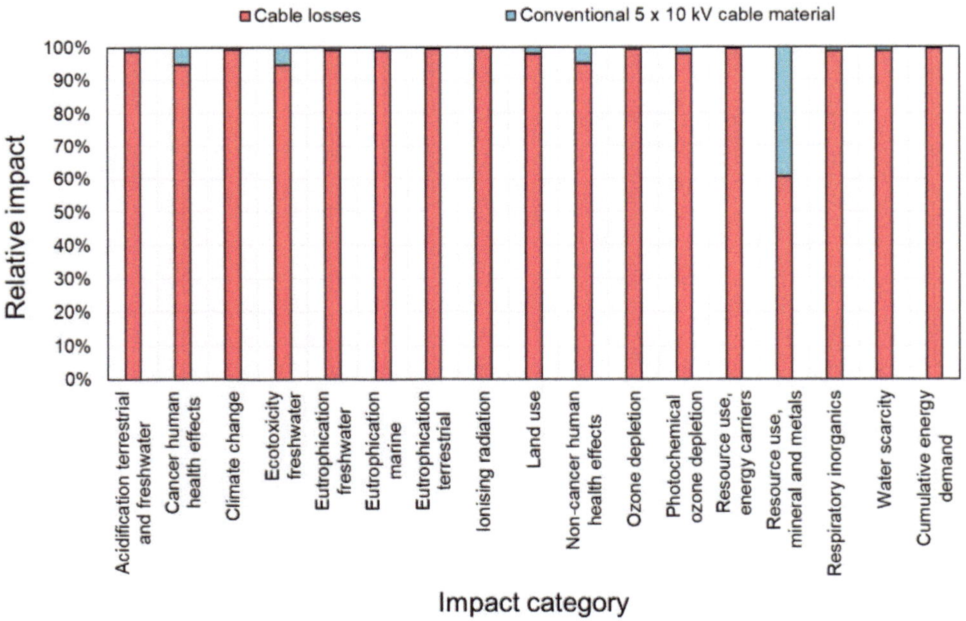

Figure 4.18: Relative impact of the use phase and the material demand of the conventional 5 x 10 kV cable system for each impact category at a load factor of $m_a = 0.7$.

4.5.1.4 Cable Impact Comparison

Comparing all three cable systems at $m_a = 0.7$ shows that the conventional 5 x 10 kV cable system is the worst alternative in 15 out of 17 impact categories. As shown in Figure 4.19, the superconducting 10 kV cable system has the highest impacts in the categories cancer human health effects and resource use of mineral and metals.

The conventional 110 kV cable system is the best out of all three cable systems in each impact category. Additionally, the impacts of the high voltage cable at most cause 16.6 % of the impacts of the respective cable system with the maximum impact in each category.

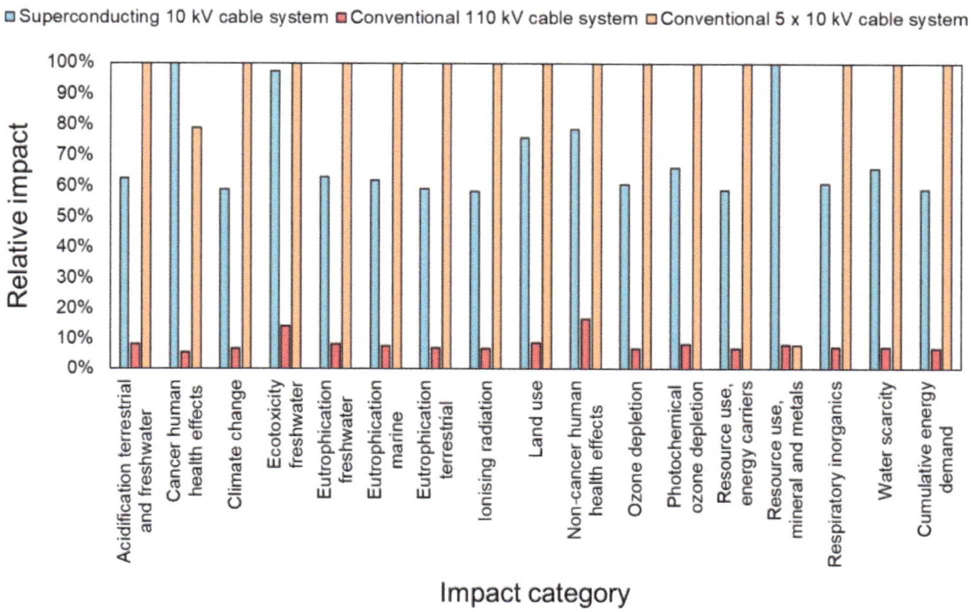

Figure 4.19: Relative impact of the three cable systems for each impact category at a load factor of $m_a = 0.7$. The cable system with the highest impact in each category is indicated by the given value of 100 %. The corresponding impact of the other cable systems is given as a relative share of 100 %.

Especially in the category resource use of mineral and metals, the superconducting cable performs much worse than both conventional cable systems. Thus, it is analysed which minerals or metals cause the highest impact for all three cable systems. Figure 4.20 shows the contribution of all minerals and metals to the total impact of each cable system at a load factor of $m_a = 0.7$. It must be noted, that the minerals and metals shown in the graph are not necessarily a component in one of the cable systems. These resources are rather extracted in one of the processes along the supply chain, even if only as a by-product as it is the case for gold. While gold has the highest share for the superconducting system, it is mostly extracted during the mining for silver and nickel and within the supply chain of copper, all of which are components of the superconducting tape. However, as gold has the highest impact factor among all mineral and metals, even smaller amounts of gold result in a high impact in the category resource use of minerals and metals. Silver has the second highest share of all metals in the impact of the superconducting cable system. As silver has the fourth highest impact factor of all resources in this category, the same effect as with gold applies where smaller amounts already cause a high impact. The impact factors of all considered metals are provided in appendix E. While copper has the third highest share in the impacts of the superconducting cable, it dominates the impacts of the conventional cable systems.

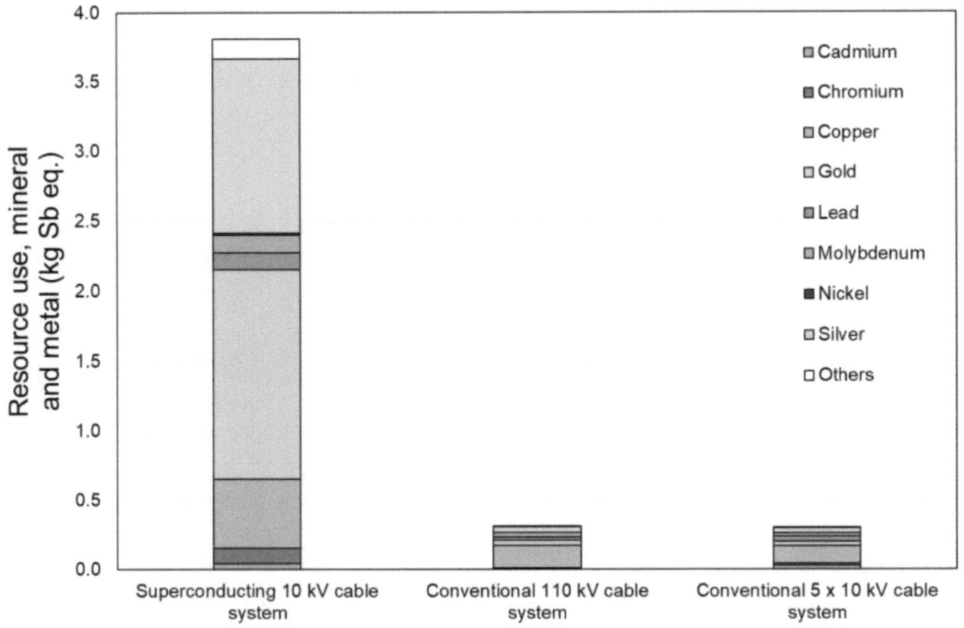

Figure 4.20: Contribution of various metals to the total impacts of each cable system in the category resource use, mineral and metal at a load factor of $m_a = 0.7$.

As one of the aims of this study is to identify the circumstances under which a superconducting cable might outperform the conventional alternatives, the impacts of all cable systems are analysed as a function of the load factor. In doing so, a break-even load factor can be identified which represents the load that is necessary for the superconducting cable to be more environmentally friendly than the conventional alternatives. However, since the impact factors used for the use phase impact calculation have an uncertainty to them this leads to an uncertainty in the results and thus in the break-even load factors.

As shown in Table 4.11 there is a range of possible impact factors for liquid nitrogen production and electricity. Figure 4.21 shows the climate change impacts for all three cable systems as a function of the load factor. Besides the baseline value, for each cable system a worst-case scenario using the highest impact factors and a best-case scenario using the lowest impact factor is shown. Instead of a single break-even load factor, this leads to a range of potential break-even load factors.

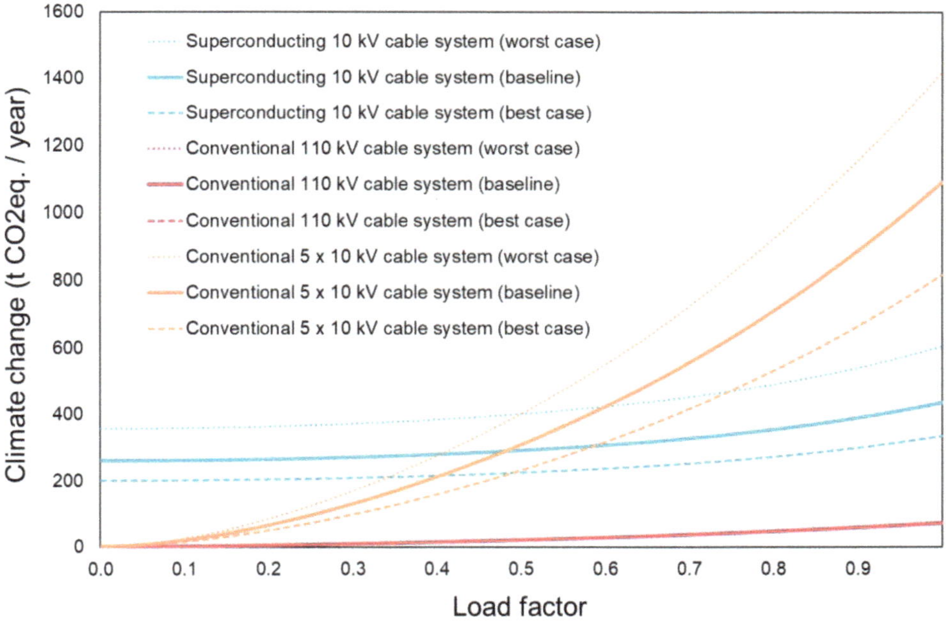

Figure 4.21: Climate change impacts of each of the three cable systems. The dashed and dotted lines indicate the best case and worst case of each cable system based on the impact factor uncertainty analysis.

When comparing the superconducting cable system with the conventional 5 x 10 kV cable system, the break-even load factor according to the baseline impact factors is at $m_a = 0.48$. However, if the best-case is assumed for the superconducting cable system and the worst case is assumed for the conventional 5 x 10 kV cable system, the break-even load factor is at $m_a = 0.34$. Vice versa, in the worst possible case the superconducting cable breakes even at $m_a = 0.75$. Thus, the break-even load factor can be described as $0.34 \leq m_a \leq 0.75$, with 0.48 being the most likely value according to the Monte Carlo simulation.

Compared with the the conventional 110 kV cable system, the superconducting cable system does never reach a break-even point. Even if the best case is assumed, the impacts of the superconducting cable are much higher than the ones of the conventional cable system.

The same comparison can now be done for each of the 17 impact categories to identify the range of potential break-even load factors. Figure 4.22 shows the ranges of potential break-even load factors for each impact category for the comparison between the superconducting cable system and the conventional medium voltage cable.

In this graph, the colour indicates which value is the most likely break-even factor with darker colours representing higher likelihoods. The size of the bar indicates the actual range of the result and thus a larger bar represents a higher uncertainty of the results. If a bar reaches the value 1, this means that in the worst possible case the superconducting cable does never reach a break-even load factor. No bar at all indicates, that the superconducting cable does never reach a break-even point even if the best possible case is assumed.

It is notable, that the superconducting cable system can reach a break-even load factor when compared to the conventional medium voltage cable for all but one impact category. In five categories, even in the worst possible case the superconducting cable reaches a break-even load factor to outperform the conventional 5 x 110 kV cables. In eleven categories, only in the worst possible case the superconducting cable does not break even with the conventional alternative.

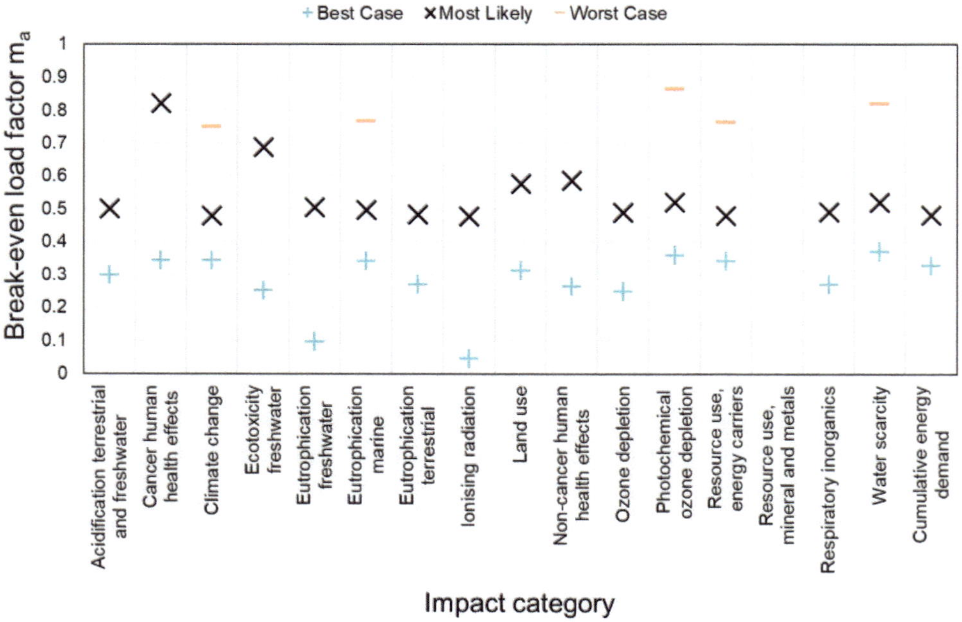

Figure 4.22: Break-even load factors for each impact category for the comparison between the superconducting 10 kV cable system and the conventional 5 x 10 kV cable system. The plus signs (+) indicate the break-even load factor for the best-case scenario, whereas the x signs represent the most likely break-even load factor and the minus signs (-) indicate the break-even load factor for the worst-case scenario. If there is no minus sign, that means in the worst case there is no break-even point. If there is no x sign, that means that most likely there is no break-even point. If there are no signs at all, that means that even in the best case there is no break-even point.

Depending on the impact category, the most likely break-even load factor is between $m_a = 0.48$ in the category ionising radiation and $m_a = 0.82$. Across all impact categories, the average most likely break-even factor is at $m_a = 0.54$. This means that on average a cable load of 54 % is enough for the superconducting cable to outperform the conventional medium voltage cable.

Considering the best possible case, the break-even load factors range from $m_a = 0.05$ to $m_a = 0.37$, with an average of 0.28 across all impact categories.

The comparison with the conventional high voltage cable shows, that simply replacing one with a superconducting medium voltage cable system will only lead to environmental disadvantages. As can be seen in Figure 4.23, the superconducting cable does never reach a break-even point with the conventional 110 kV cable system in 14 out of 17 categories. In the remaining three categories, the most likely case is still that no break-even point can be reached.

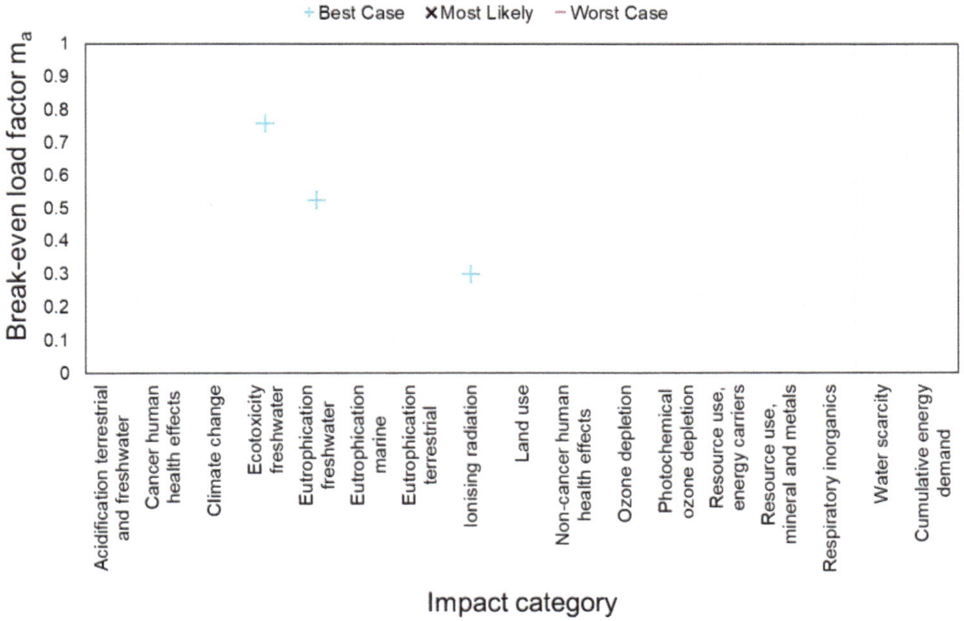

Figure 4.23: Break-even load factors for each impact category for the comparison between the superconducting 10 kV cable system and the conventional 110 kV cable system. The plus signs (+) indicate the break-even load factor for the best-case scenario, whereas the x signs represent the most likely break-even load factor and the minus signs (-) indicate the break-even load factor for the worst-case scenario. If there is no minus sign, that means in the worst case there is no break-even point. If there is no x sign, that means that most likely there is no break-even point. If there are no signs at all, that means that even in the best case there is no break-even point.

However, in the best possible case, the superconducting cable can reach a break-even load factor. Nevertheless, with values of 0.76 (ecotoxicity freshwater), 0.53 (eutrophication freshwater) and 0.30 (ionising radiation) for m_a these break-even points are quite high for a best possible case. Additionally, the best possible case is also a rather unlikely one and thus it can be assumed that a superconducting medium voltage cable cannot outperform a conventional high voltage cable. At least not, if the system boundaries are as shown in Figure 4.2.

4.5.2 Superconducting Medium Voltage Cable vs. Conventional High Voltage Cable with Changed Transformer Configuration

A superconducting medium voltage cable system is not an alternative for replacing conventional high voltage cables in inner city areas in terms of environmental impacts. However, the aim of this study is to identify if there are any conditions under which a superconducting cable system can perform better than a conventional 110 kV cable.

For this reason, changes are made to the original system boundaries. Figure 4.24 shows the system boundaries of a scenario comparison. In this comparison, the transformer configuration of the superconducting cable system is changed by replacing the 380/110 kV transformer with a 380/10 kV transformer and thus getting rid of the second transformer. As the superconducting cable system now only has one transformer, the second transformer in the conventional cable system must be included in the system boundaries to account for the additional losses and material demand.

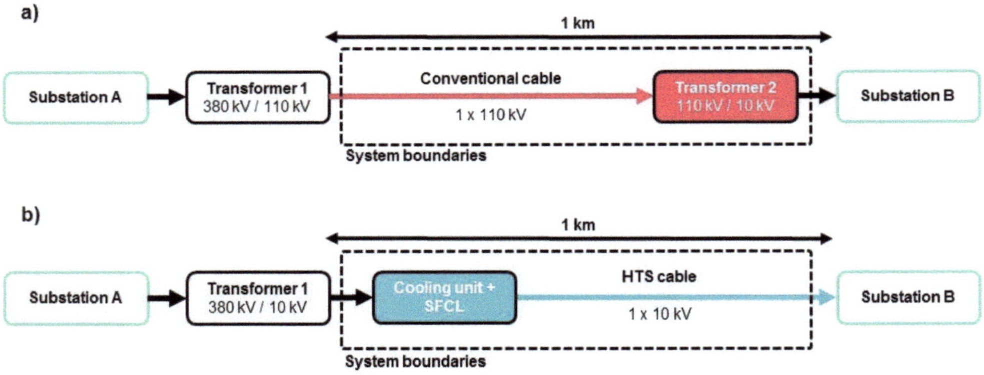

Figure 4.24: System boundaries for the scenario options of a) a 110 kV conventional cable and the required second transformer and b) a high-temperature superconducting (HTS) cable and its required cooling system (based on [163]).

For this comparison, the assumption is made that the first transformers in both systems have similar losses and can thus be neglected. As the second transformer is now included in the conventional cable system, the unit processes that must be considered in the comparison changed. Figure 4.25 shows the adjusted detailed system boundaries for the conventional 110 kV cable system including the unit processes for the transformer production.

In addition to the increased material demand, the total losses of the system also change as the transformer introduced additional losses. These transformer losses P_{trans} consist of the ohmic losses of the two copper coils due to the winding resistances and the load-independent iron core losses P_{Fe}. The transformer losses are calculated according to equation 13.

$$P_{trans} = 3 * ({I_U}^2 * R_U + {I_L}^2 * R_L) + P_{Fe} \quad \text{(Eq. 13)}$$

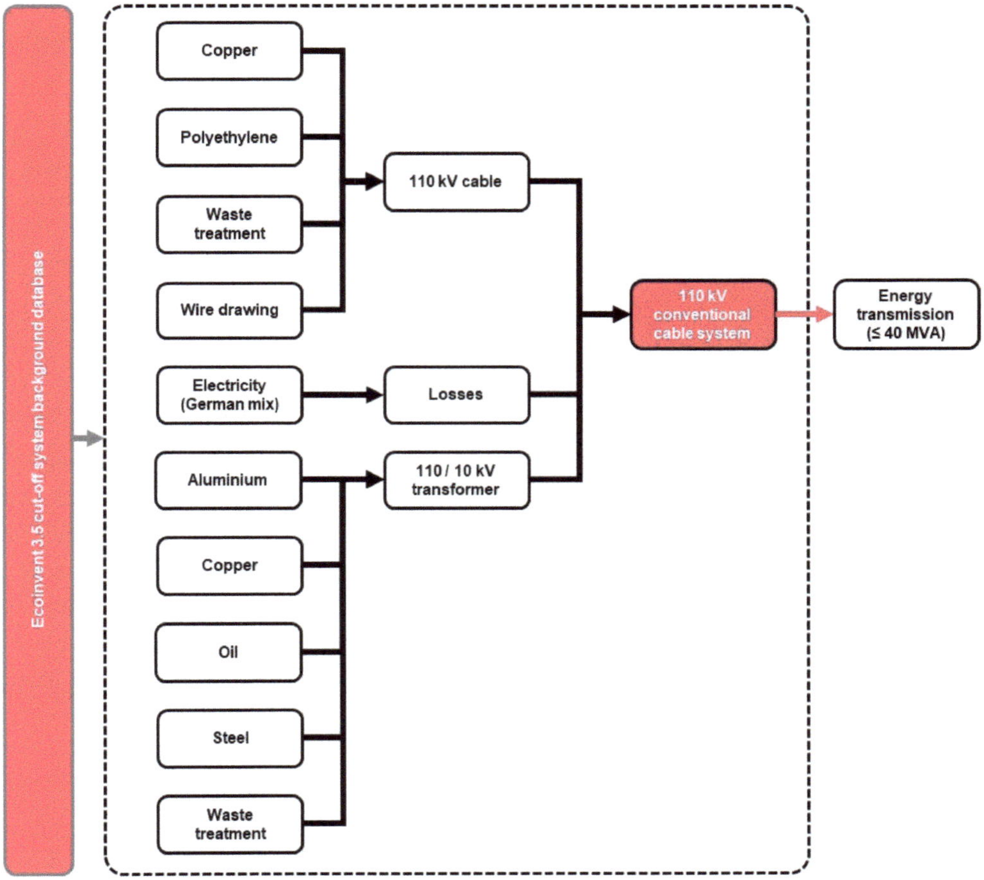

Figure 4.25: System boundaries of the 110 kVconventional cable system in a scenario comparison with an included transformer.

In this equation, the indices U and L denote the respective nominal current I and resistance R of the coil of the upper voltage level (U) and the coil of the lower voltage level (L). For the respective resistances, values of a 63 MVA transformer are taken from literature. These resistances are 39.2 mΩ for the upper coil and 5.6 mΩ for the lower coil [171]. Although this transformer has a higher rated power than the 40 MVA that are used for the functional unit of this study, the transformer is similar in terms of rated currents. For this reason, the approximation is considered sufficient for the loss calculation. The iron core losses are composed of the iron loss factor and the mass of the iron core. For the selected transformer, the iron loss factor is 1.13 kW/t and the mass of the iron core is 21.2 t. This results in load-independent iron core losses of about 24 kW. Table 4.12 shows the hourly losses of the conventional 110 kV cable as well as the transformer losses and the resulting total system losses for various cable loads. For comparison, the total system losses of the superconducting 10 kV system are also shown. It can be seen, that the included transformer drastically increases the total conventional cable system losses, making them even higher than the superconducting cable system losses at higher loads.

Table 4.12: Hourly losses of the conventional 110 kV cable system including transformer losses. For comparison, the total system losses of the superconducting system consisting of the electricity consumption of the liquid nitrogen production and the pumps are also shown. All values are given in kW.

	10% S_n	**30% S_n**	**50% S_n**	**70% S_n**	**90% S_n**	**S_n = 40 MVA**
Losses of the conventional 110 kV cable $P_{conv, 110}$	0.13	1.14	3.16	6.19	10.23	12.63
Losses of the transformer P_{trans}	24.90	32.49	47.65	70.40	100.73	118.74
Total losses of the conventional system, $P_{total, conv}$	25.03	33.62	50.81	76.59	110.96	131.37
HTS total system losses P_{total}	43.31	44.13	46.88	52.98	63.78	71.36

Figure 4.26 shows the resulting annual system losses of the conventional 110 kV system including the transformer losses in direct comparison with the superconducting 10 kV cable system for all load factors. This comparison shows that the system losses of the superconducting cable system with the adjusted transformer configuration are lower than the total conventional cable system losses for a load factor of $m_a \geq 0.43$. From an energy-related point of view, this transformer configuration makes it possible for a superconducting medium voltage cable to replace a conventional high voltage cable in inner-city areas while simultaneously saving space due to less transformers.

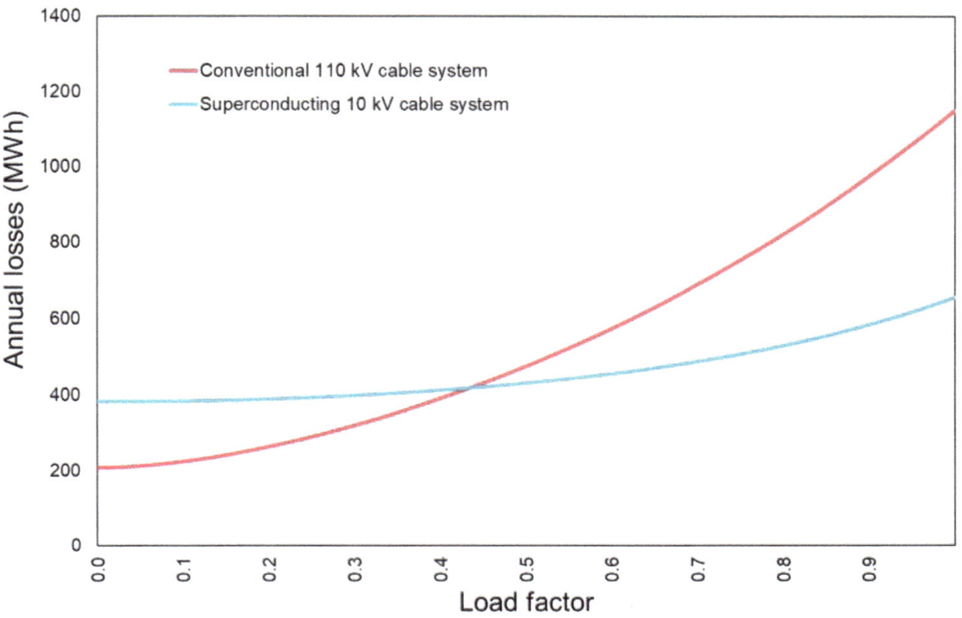

Figure 4.26: Annual loss energy as a function of the load factor for each of conventional high voltage system including a transformer and the superconducting medium voltage system.

For a complete comparison however, not only the use phase must be considered. Because of the changed system boundaries, the material demand for the transformer must be considered as well. For this purpose, literature values from ABB are used, which provides a detailed list of materials required for a 40 MVA transformer within an environmental product declaration [169]. The complete list of materials as presented in this environmental product declaration is given in Table 4.13.

Table 4.13: Life cycle inventory of a 40 MVA transformer [169].

Material	Input Amount	Unit
Aluminium	92.8	kg
Brass	40.8	km
Copper	9.0	t
Epoxy resin	6.0	kg
Glass fiber	462.0	kg
Kraft paper	420.8	kg
Oil	15.5	t
Paint	38.0	kg
Steel	35.8	t
Waste treatment aluminium	-92.8	kg
Waste treatment copper	-9.0	t
Waste treatment steel	-35.8	t
Material	**Output Amount**	**Unit**
40 MVA transformerr	1	item

4.5.2.1 Contribution Analysis for the 110 kV Conventional Cable System

In the climate change category, the use phase accounts for the largest share of total impacts, as shown in Figure 4.27. Even at low loads, the losses of the transformer cause more than 95.0 % of the total impacts due to the iron-core losses. The total share of the cable and transformer losses increase to 99.4 % with increasing load.

Material-wise, the transformer has a higher share than the conventional 110 kV cable. However, compared to the use phase impacts these material impacts are negligible.

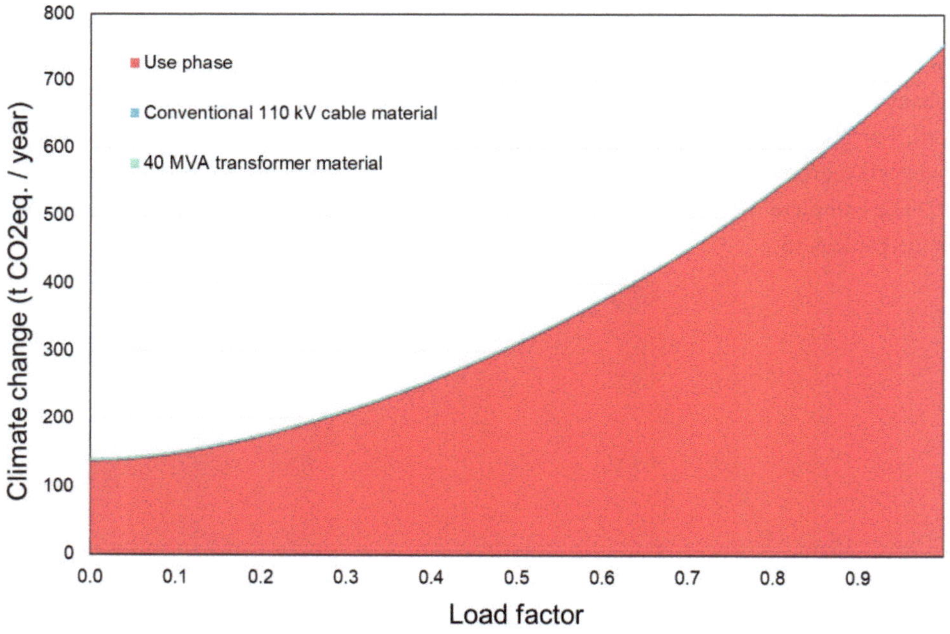

Figure 4.27: Climate change impacts by source as a function of load factor for the conventional 110 kV cable system including a 40 MVA transformer.

At a load factor of 0.7, the impacts in all categories are dominated by the transformer losses as can be seen in Figure 4.28. On average, the transformer losses cause 81.6 % of the total impacts. The losses of the cable have a mean share of 7.3 % across all impact categories. Only in the category resource use of mineral and metals, the material demand has a higher share than the use phase. The material demand for the transformer losses cause 51.6 % of the total impact, while the material demand for the conventional 110 kV cable cause 37.6 % of the resource use impacts.

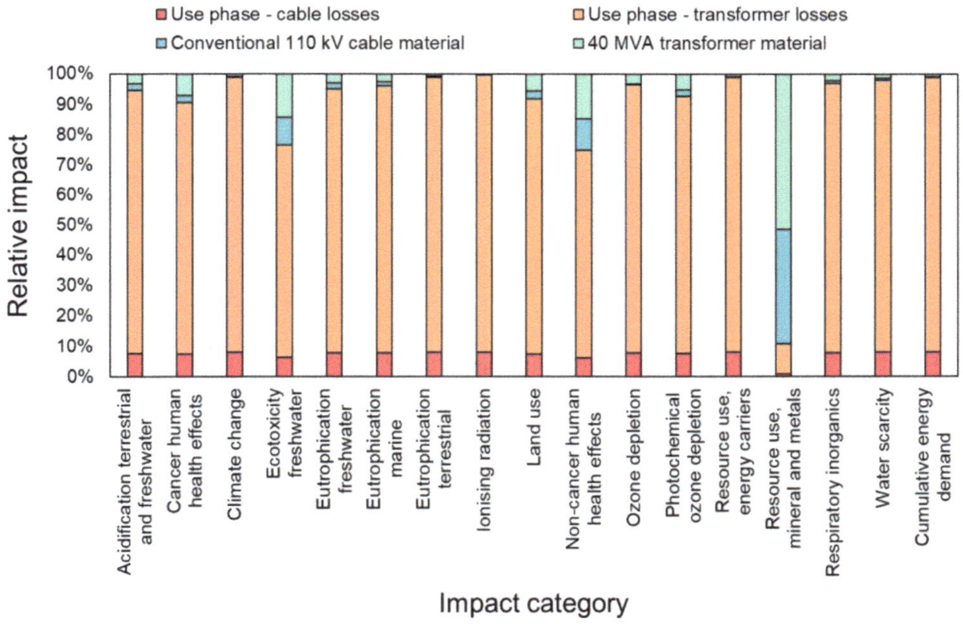

Figure 4.28: Relative impact of the use phase and the material demand of the conventional 110 kV cable system including a transformer for each impact category at a load factor of $m_a = 0.7$.

4.5.2.2 Cable Impact Comparison with Changed Transformer Configuration

Comparing all impact categories shows that the superconducting 10 kV cable system performs better than the conventional high voltage alternative in 14 out of 17 impact categories when the transformer configuration is changed, as can be seen in Figure 4.29.

The superconducting cable system only performs worse in the categories cancer human health effects, ecotoxicity freshwater and resource use of mineral and metals. In the latter category, the impacts of the conventional cable system are only about one fifth of the impacts of the superconducting cable system. In the categories in which the superconducting cable system performs better, its impacts are about three quarters of the conventional cable system impacts on average.

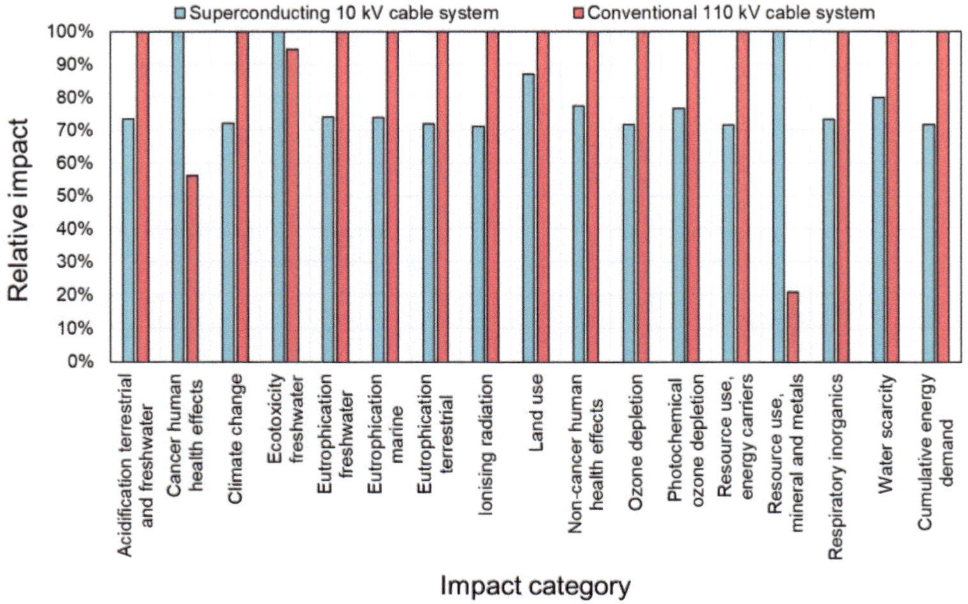

Figure 4.29: Relative impact comparison of the superconducting 10 kV cable system and the conventional 110 kV cable system including a 40 MVA transformer for each impact category at a load factor of $m_a = 0.7$. The cable system with the higher impact in each category is indicated by the given value of 100 %. The corresponding impact of the other cable system is given as a relative share of 100 %.

As the superconducting cable system does perform better in most impact categories, it follows that there must be a break-even load factor between the superconducting cable system and the conventional 110 kV cable system.

Figure 4.30 shows the climate change impacts for both cable systems as a function of the load factor including the results of the uncertainty analysis for both. While there was no break-even load factor with the original transformer configuration, in the adjusted system the most likely break-even load factor is at $m_a = 0.45$. Considering the best possible case, the superconducting cable breaks even at $m_a = 0.26$, while in the worst possible case the break-even load factor is $m_a = 0.71$. This means, that with the considered transformer configuration, the superconducting cabel system will perform better than the conventional cable system as long as the load is sufficiently high.

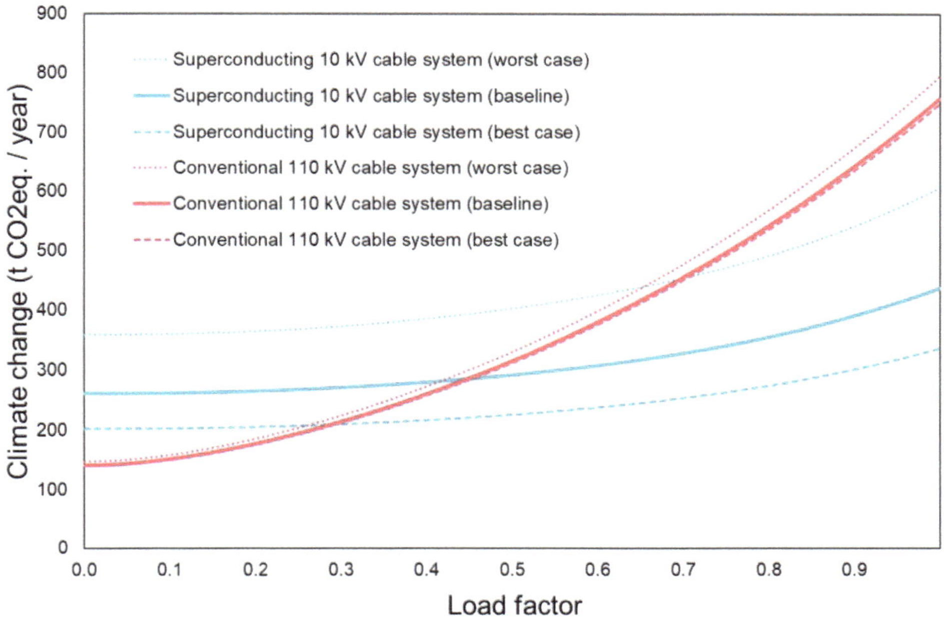

Figure 4.30: Climate change impacts of each of the superconducting 10 kV cable system and the conventional 110 kV cable system including a 40 MVA transformer. The dashed and dotted lines indicate the best case and worst case of each cable system based on the impact factor uncertainty analysis.

The same analysis is done for each of the impact categories as shown in Figure 4.31. For all impact categories, there is a potential break-even load factor between the superconducting 10 kV cable system and the conventional 110 kV cable system. The only exception is the impact category resource use of minerals and metals, in which no break-even point can be reached. In the category cancer human health effects, most likely no break-even point can be reached as well. Nevertheless, in this category the superconducting cable system can break even in the best possible case even at no load. This result, however, shows the high uncertainty of the results in this particular impact category.

In all other categories, the average most likely break-even load factor is at $m_a = 0.48$. It must be mentioned that for 13 of the 17 impact categories in the worst possible case no break-even point can be achieved.

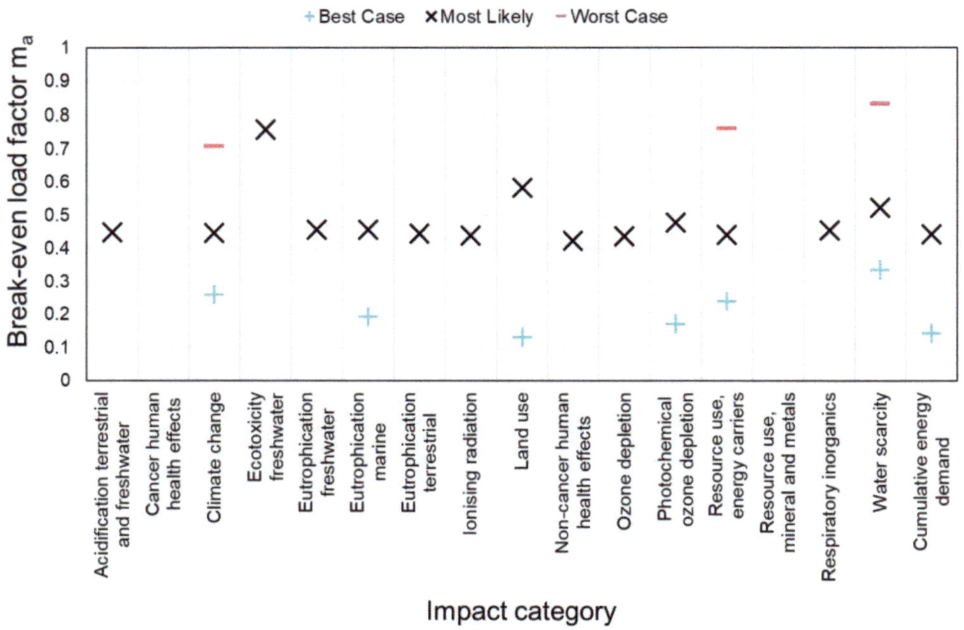

Figure 4.31: Break-even load factors for each impact category for the comparison between the superconducting 10 kV cable system and the conventional 110 kV cable system including a 40 MVA transformer. The plus signs (+) indicate the break-even load factor for the best-case scenario, whereas the x signs represent the most likely break-even load factor and the minus signs (-) indicate the break-even load factor for the worst-case scenario. If there is no minus sign, that means in the worst case there is no break-even point. If there is no x sign, that means that most likely there is no break-even point. If there are no signs at all, that means that even in the best case there is no break-even point. If there are no plus signs but other signs, that means that in the best case, the superconducting cable performs better even at no load.

On the contrary, in the best possible case the superconducting cable will perform better in nine impact categories even if there is no cable load. Thus, changing the transformer configuration from two transformers to one 380/10 kV transformer is a potential option to make a superconducting medium voltage cable system perform better than a conventional high voltage system. Therefore, a superconducting 10 kV cable system can potentially replace conventional 110 kV cables when using the considered transformer configuration.

4.5.3 Open Cooling System vs. Closed Cooling System

In addition to changing the transformer configuration, there are other potential changes that can be made to the system to analyse if the superconducting system can outperform a conventional high voltage system. As most of the environmental impacts of the superconducting cable system can be traced back to the production of the necessary liquid nitrogen, this study also consideres the use of a closed cooling system. Such a closed cooling system does not lose liquid nitrogen over time as it uses a cryocooler to electrically re-cool the heated liquid nitrogen. However, while a closed cooling system does not require constant refilling, the cryocooler causes a higher electricity consumption. Therefore, the first step that has to be made is to compare the impacts of the open cooling system to the closed cooling system.

4.5.3.1 Closed Cooling System Components

Because this study is based on the AmpaCity cable in Essen, Germany with its open cooling system, a hypothetical closed cooling system must be considered for this comparison. The superconducting cable system in Albany, USA, is similar to the AmpaCity cable and uses a closed cooling system [172]. Therefore, the hypothetical closed cooling system in this study is modelled after this cooling system.

A schematic representation of the Albany cooling system is shown in Figure 4.32. Technically, this cooling system must be considered a hybrid cooling system rather than a closed cooling system. The reason for this is that it uses components of an open cooling system for redundancy reasons. Thus, in the event of a failure the system can switch into an open system mode. However, in this study a failure is not considered and continuous operation as a closed cooling system is assumed. Nevertheless, the material demand for all redundancy components is included in this study.

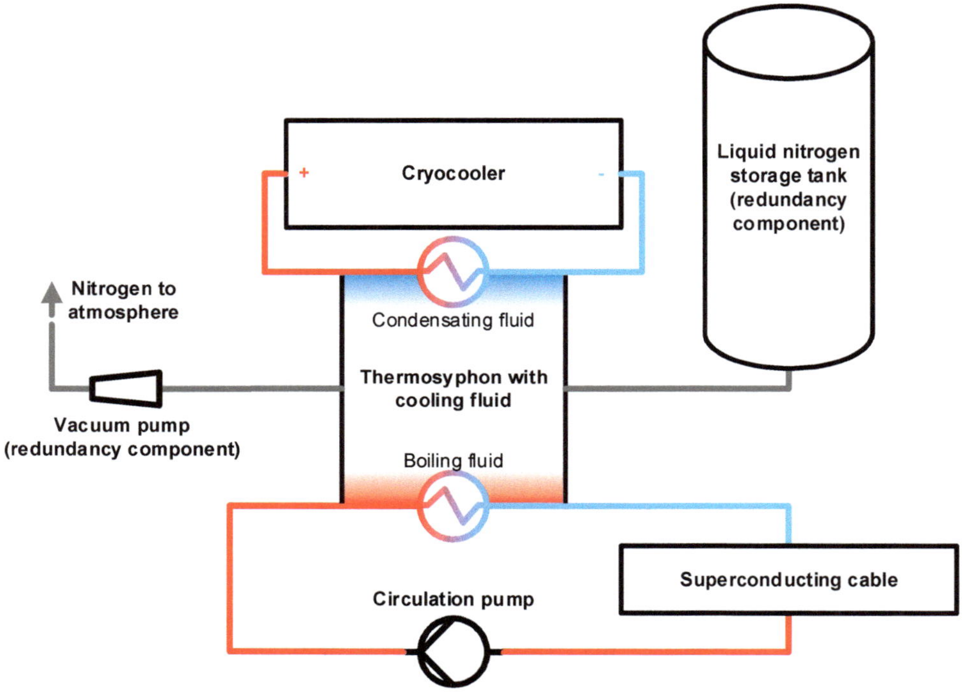

Figure 4.32: Schematic representation of a closed cooling system based on the cooling system of the superconducting cable in Albany, NY, USA [172].

In this system, a Stirling cryocooler with a cooling efficiency of $\eta = 10\ \%$ is assumed for the cooling system. This means that ten watts of electrical energy at room temperature are required for each watt of cooling capacity at low temperature. The cooling capacity is provided here via the compression and expansion of an internal cooling liquid such as helium. The cryocooler requires an additional storage tank. While this storage tank is assumed to be similar to the storage tank of the open cooling system, its capacity is only about 1/7 of the capacity of a storage tank of an open cooling system. Detailed life cycle inventories of each component specific to the closed cooling system are provided in appendix F. The detailed system boundaries are shown in Figure 4.33.

The cooling energy is provided via a thermosyphon that contains two heat exchangers and a cooling fluid. One heat exchanger is connected to the cooling circuit of the superconducting cable. The cooling fluid that is inside the thermosyphon absorbs the heat from the liquid nitrogen, which is circulated through the superconducting cable, and starts to boil and evaporate. The second heat exchanger is connected to the cryocooler and re-cools the evaporating cooling fluid causing it to condensate again. In liquid form, the cooling fluid can again absorb heat from the liquid nitrogen and the process starts from the beginning.

As with the open cooling system, a lifetime of 40 years is assumed for all components of the closed cooling system. The life cycle inventories for the individual components are taken from the study by Jacob and are provided in annex E [167].

Due to the changed cooling system, the calculation of the total system losses of the use phase also changes. While equations 1-8 are still valid for this system, the total system losses do not depend on the amount and production of liquid nitrogen. Instead, the cooling efficiency η of the cryocooler must be considered. As with the open system, the current-dependent losses and the voltage-dependent losses must be considered as grid losses and as thermal input into the circulating liquid nitrogen. In addition, the thermal losses and the pump heat impact also cause a heat input into the liquid nitrogen. The total heat input causes an electricity consumption of the cryocooler based on its efficiency η. Thus, the total system losses of the closed cooling system $P_{total,closed}$ can be calculated using equation 14.

$$P_{total,closed} = \frac{1}{\eta} * P_{Thermal} + \left(1 + \frac{1}{\eta}\right)\left[P_{Volt} + P_{Current,I_c} * \left(\frac{I}{I_c}\right)^3\right] + \frac{1}{\eta} * P_{Qp} \quad \text{(Eq. 14)}$$

In general, the total system losses of the closed cooling system are lower as can be seen from Table 4.14. At full load, the total heat input into the system is 6.2 kW in one hour. For the open cooling system, this leads to total system losses of 71.4 kW in an hour due to the liquid nitrogen production. As the closed cooling unit uses a more efficient cryocooler, the total losses are only 64.4 kW in one hour.

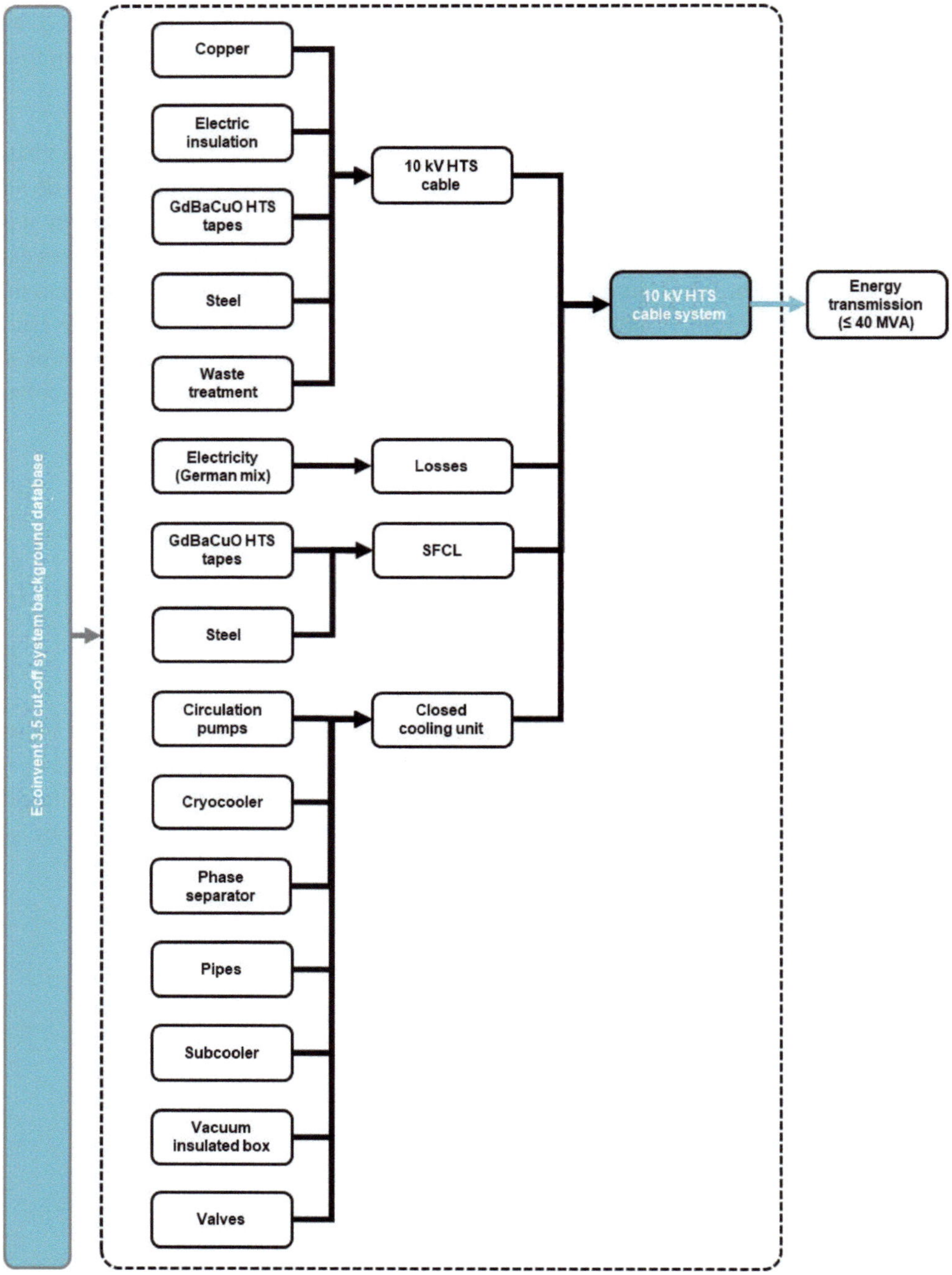

Figure 4.33: System boundaries of the 10 kV high-temperature superconducting cable system with a closed cooling unit and a superconducting fault current limiter.

Table 4.14: Comparison of the total system losses of the superconducting cable system using either an open or a closed cooling system. For reference, the total heat input that requires cooling is also given. All values are given in kW.

	10% S_n	30% S_n	50% S_n	70% S_n	90% S_n	S_n = 40 MVA
Total heat input P_{Qtotal}	3.39	3.47	3.74	4.34	5.40	6.15
Total HTS system losses (open cooling system)	43.31	44.13	46.88	52.98	63.78	71.36
Total HTS system losses (closed cooling system)	34.05	34.84	37.81	44.43	56.16	64.39

This difference is reflected in the annual loss energy of both systems. As shown in Figure 4.34 the general trend of the loss profile is the same, however the closed cooling system consumes up to 54 7 MWh less energy per year.

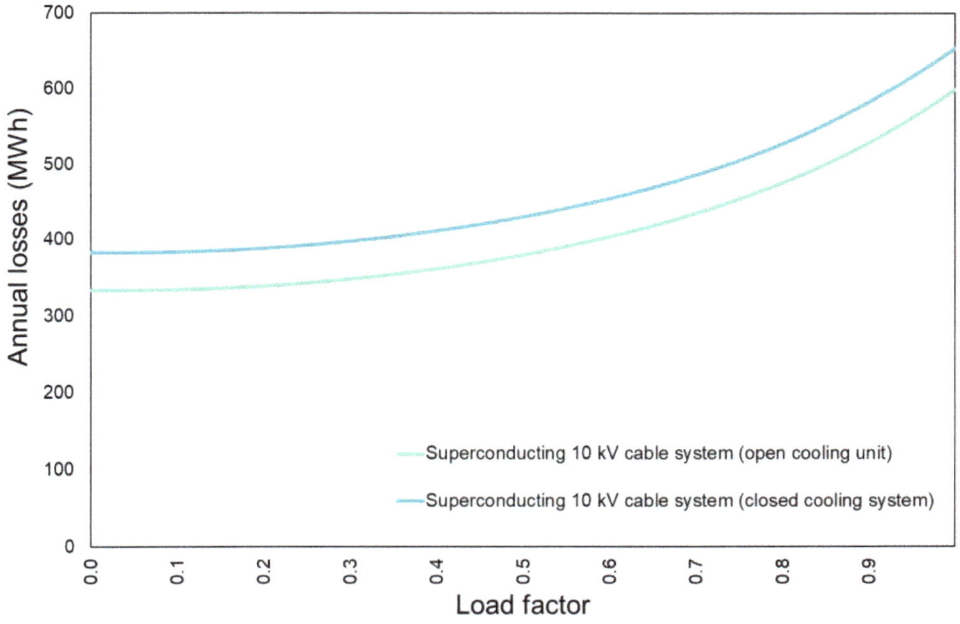

Figure 4.34: Comparison of the annual loss energy of the superconducting 10 kV cable system with either a closed or an open cooling system.

4.5.3.2 Contribution Analysis for the Closed Cooling System

For the closed cooling system, the average impact share of the use phase across all impact categories is 77.7 % and thus slightly less than the share of the use phase for the open cooling system. This is because the material demand for the closed cooling unit is slightly higher while also the total system losses are decreased. The share of the use phase as well as the shares of the superconducting cable, the fault current limiter, and the closed cooling unit are shown in Figure 4.35. The use phase dominates all categories except for the impact categories cancer human health effects and resource use of mineral and metals. In both categories, the material demand has a share of more than 50 %, while in the latter the superconducting cable alone accounts for 88.4 % of the total impact. On average, the share of the closed cooling unit is only 4.0 % across all categories. However, in the categories cancer human health effects and ecotoxicity freshwater, the share of the cooling system is 25.1 % and 13.7 % respectively.

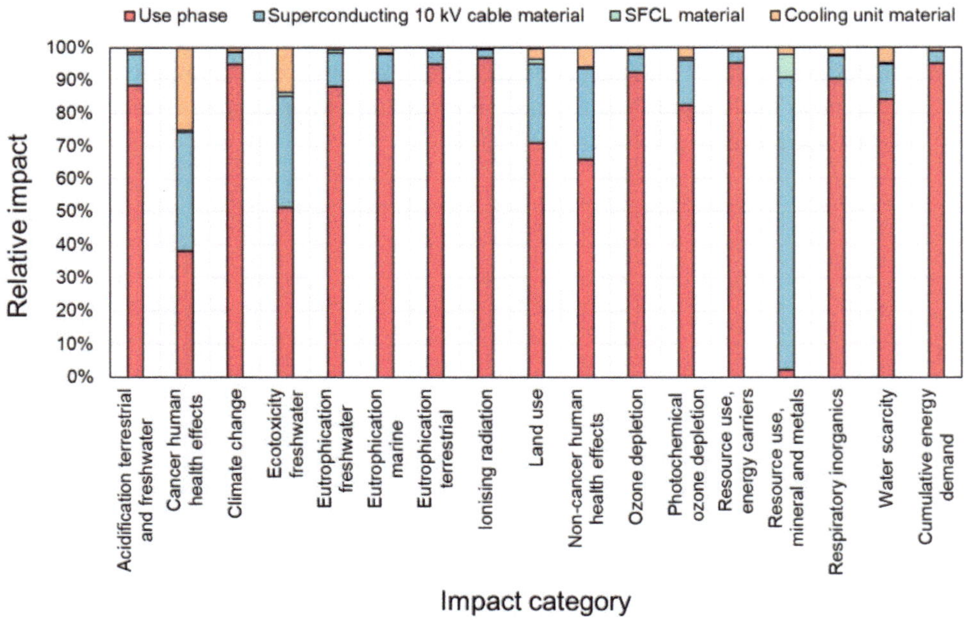

Figure 4.35: Comparison of the use of different cooling units for the superconducting 10 kV cable in the category cancer human health effects.

Within the cooling unit, the liquid nitrogen storage tank has the highest share across all impact categories as is shown in Figure 4.36. On average, it is responsible for 67.2 % of the cooling unit impacts. This means that the highest impact of the material demand is caused by a component which is primarily used for reasons of redundancy. The second highest share in the impacts of the cooling unit materials is caused by the cryocooler. On average, the cryocooler impacts account for 17.9 % of the total impacts. The thermosyphon, the circulation pumps as well as the miscellaneous components all have a combined average impact share of less than 15.0 %.

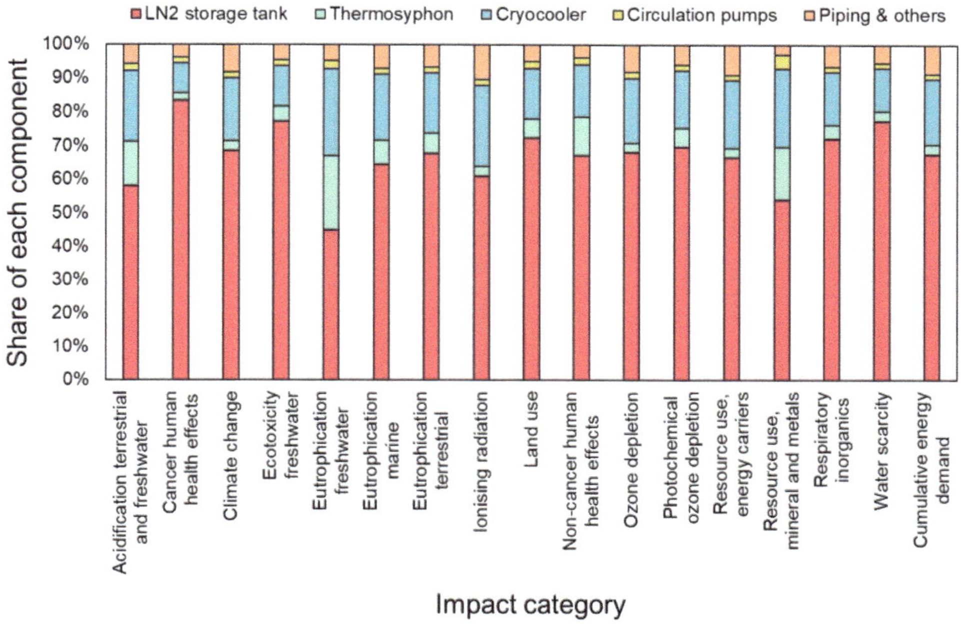

Figure 4.36: Share of each component of the superconducting cable system with a closed cooling unit to the total impact of the material demand for a load factor of $m_a = 0.7$.

4.5.3.3 Cooling System Impact Comparison

As the closed cooling system has lower system losses but higher material demand, it is necessary to compare the environmental impacts of both systems over their entire life cycle. Figure 4.37 shows the climate change impacts of both cooling systems as a function of the load factor. As with the total system losses, the climate change impacts of the closed cooling system are lower regardless of the load. At lower loads, using a closed cooling system results in a greenhouse gas emission reduction of 12.0 %, while at full load the reduction is still at 8.0 %.

This means that in terms of climate change impact, the closed cooling system is always the preferable option for a 1 km long, 10 kV, 40 MVA superconducting cable system.

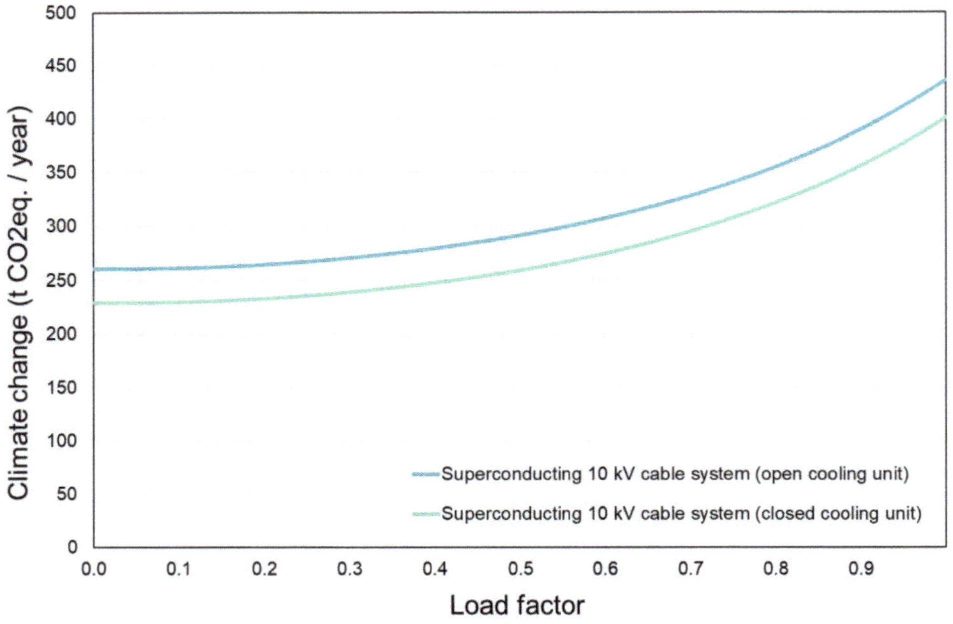

Figure 4.37: Comparison of the climate change impacts of the different cooling systems for the superconducting 10 kV cable system.

When comparing all impact categories at a load factor of $m_a = 0.7$, it shows that the closed cooling system performs better in each of the 17 impact categories, as is shown in Figure 4.38.

On average, the environmental impacts are reduced by 10.0 % when using a closed cooling system instead of an open cooling system. In the category resource use of mineral and metals, an impact reduction of even 23.9 % is achieved. Only in the category ecotoxicity freshwater, the impact savings are below 5 % with a savings value of 3.9 %.

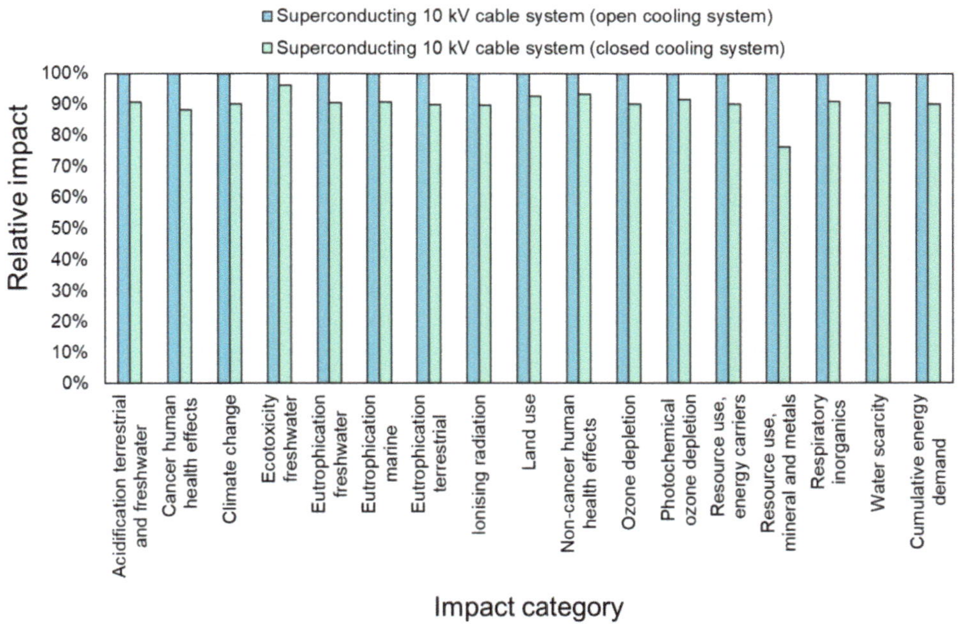

Figure 4.38: Relative impact comparison of the superconducting 10 kV cable system with an open cooling system and the superconducting 10 kV cable system with a closed cooling system for each impact category at a load factor of $m_a = 0.7$. The cable system with the higher impact in each category is indicated by the given value of 100 %. The corresponding impact of the other cable system is given as a relative share of 100 %.

However, the impact savings are rather low and thus the superconducting 10 kV cable system cannot simply replace a conventional 110 kV cable system even when using a closed cooling system. Figure 4.39 shows the potential break-even load factors when comparing a superconducting medium voltage cable system with a conventional high voltage cable system with the original transformer configuration. As the graph shows, in 14 out of 17 impact categories the superconducting cable will never break even with the conventional 110 kV cable. In the other three categories, a break-even load factor is only achieved assuming a best possible case and is thus rather unlikely.

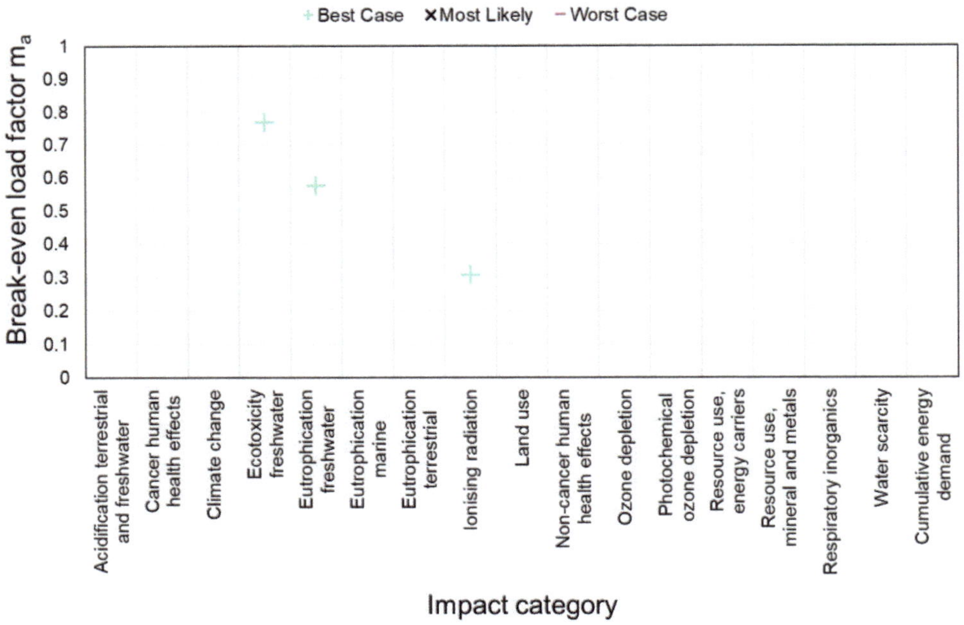

Figure 4.39: Break-even load factors for each impact category for the comparison between the superconducting 10 kV cable system with a closed cooling system and the conventional 110 kV cable system. The plus signs (+) indicate the break-even load factor for the best-case scenario, whereas the x signs represent the most likely break-even load factor and the minus signs (-) indicate the break-even load factor for the worst-case scenario. If there is no minus sign, that means in the worst case there is no break-even point. If there is no x sign, that means that most likely there is no break-even point. If there are no signs at all, that means that even in the best case there is no break-even point.

4.5.3.4 Cable Impact Comparison with Changed Transformer Configuration and Closed Cooling System Usage

Changing the transformer configuration results in the superconducting cable system being able to outperform the conventional high voltage cable system. Simply changing the cooling system does not have the same effect. Nevertheless, a combination of both approaches further improves the superconducting cable system performance when compared to the conventional cable system.

Figure 4.40 shows the climate change impacts of the superconducting cable system with a closed cooling system and a changed transformer configuration in comparison with the conventional 110 kV cable system. While the most-likely break-even load factor is at $m_a = 0.45$ when using an open cooling unit, using a closed one further reduces this value to $m_a = 0.37$.

When considering the data uncertainty, in the best possible case the superconducting cable can outperform the superconducting cable for every load factor above $m_a = 0.17$. In the worst possible case, the break-even load factor is at $m_a = 0.55$.

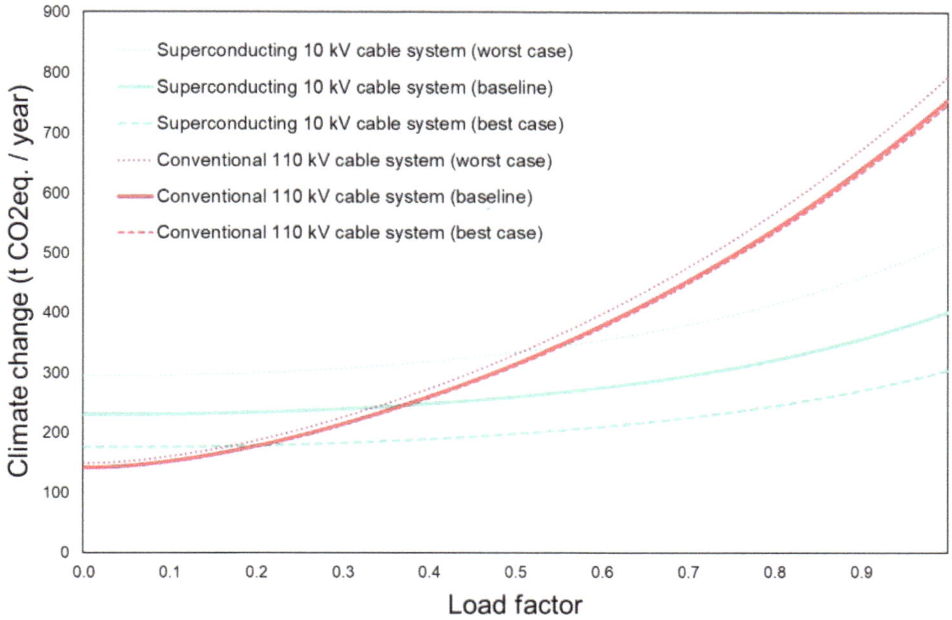

Figure 4.40: Climate change impacts of each of the superconducting 10 kV cable system with a closed cooling system and the conventional 110 kV cable system including a 40 MVA transformer. The dashed and dotted lines indicate the best case and worst case of each cable system based on the impact factor uncertainty analysis.

At aload factor of 0.7, the superconducting cable system shows a better environmental performance in 14 out of 17 categories when using a closed cooling system and the adjusted transformer configuration. The comparison of both systems is shown in Figure 4.41. In those categories where the superconducting cable system performs better, an average impact reduction of 32.2 % can be achieved.

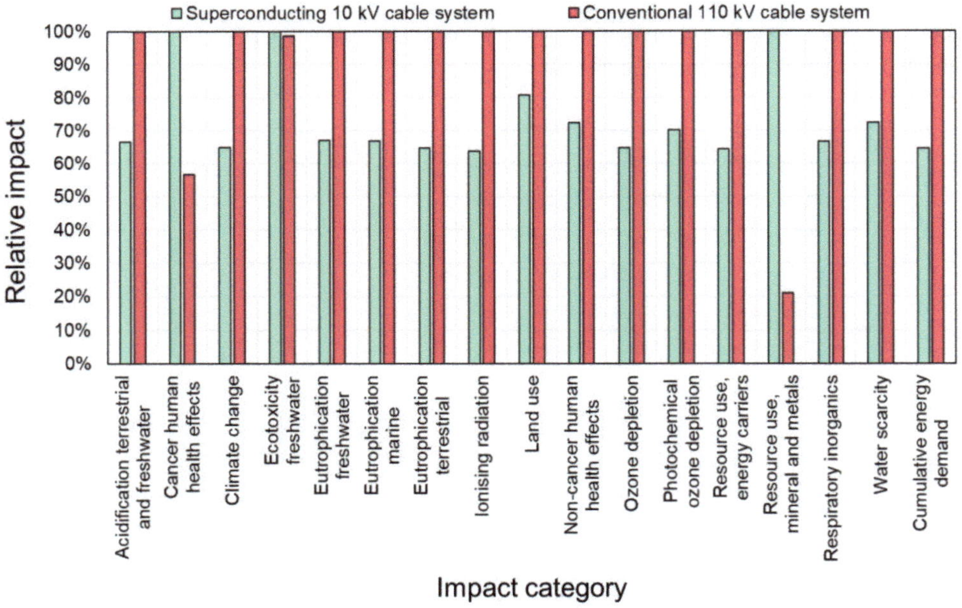

Figure 4.41: Relative impact comparison of the superconducting 10 kV cable system with a closed cooling system and the conventional 110 kV cable system including a 40 MVA transformer for each impact category at a load factor of $m_a = 0.7$. The cable system with the higher impact in each category is indicated by the given value of 100 %. The corresponding impact of the other cable system is given as a relative share of 100 %.

As shown in Figure 4.42, using a closed cooling system in addition to the changed transformer configuration further decreases the break-even load factors between the superconducting and the conventional 110 kV cable system. On average, the most likely break-even load factor is at $m_a = 0.41$. This means that even when the cable load is only at 50 %, the superconducting cable system has a high probability of being the environmentally advantegous option compared to conventional high voltage cables.

Nevertheless, in the category resource use of mineral and metals the superconducting cable can still not perform better than the conventional cable. This is the case even with both, the closed cooling unit and the adjusted transformer configuration.

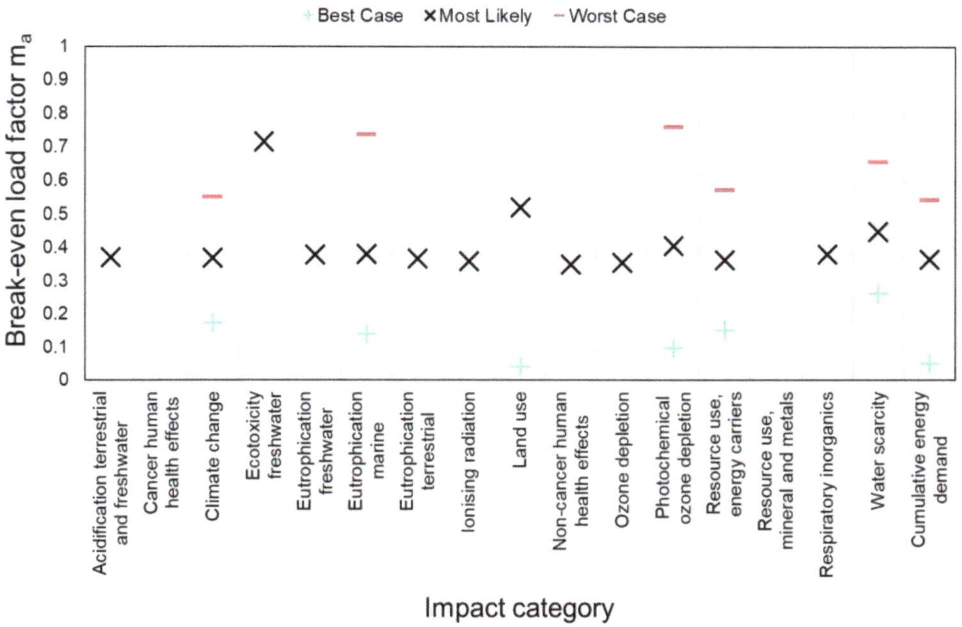

Figure 4.42: Break-even load factors for each impact category for the comparison between the superconducting 10 kV cable system with a closed cooling system and the conventional 110 kV cable system including a 40 MVA transformer. The plus signs (+) indicate the break-even load factor for the best-case scenario, whereas the x signs represent the most likely break-even load factor and the minus signs (-) indicate the break-even load factor for the worst-case scenario. If there is no minus sign, that means in the worst case there is no break-even point. If there is no x sign, that means that most likely there is no break-even point. If there are no signs at all, that means that even in the best case there is no break-even point. If there are no plus signs but other signs, that means that in the best case, the superconducting cable performs better even at no load.

Additionally, due to the data uncertainty in the worst possible case the superconducting cable will not perform better in ten impact categories. Then again, in the best possible case the superconducting cable will always be the better option in the same ten categories. This result shows that there is a high uncertainty for these ten impact categories. Nevertheless, the average most likely break-even factor for these categories is still at $m_a = 0.41$. This indicates that although there is a high uncertainty, the most likely result is still that the superconducting cable system will perform better at least for sufficiently high loads, meaning a load factor of $m_a \geq 0.41$.

4.5.3.5 Cooling System Impact Comparison with a Renewable Energy Mix

So far, the results show that the superconducting cable system can be an environmentally friendly option compared to both, a conventional medium voltage and a conventional high voltage cable system. Additionally, it is shown that the closed cooling system is the preferrable cooling system.

However, this study also aims at identifying further savings potential. In the first comparison, the cryocooler and the liquid nitrogen production are modelled to use the German electricity mix from the ecoinvent database. This electricity mix has a high share of fossil fuels. Thus, in a further scenario comparison the German electricity mix is replaced by an electricity mix that is entirely based on renewable energy.

For this reason, a renewable electricity mix from 2020 by a German provider is selected. The composition of this mix is shown in Figure 4.43. More than half of this energy mix comes from wind energy, while the rest comes from biogas, solar energy, hydro energy, and geothermal energy.

For the closed cooling system, it is assumed that the cryocooler is using this renewable electricity mix. In case of the open cooling system, it is assumed that the air separation unit is using this electricity mix to produce the liquid nitrogen. This assumption also is a limitation to this study as it implies that the operator of the superconducting cable has an influence on the electricity mix used by the air separation unit which most likely will not be the case.

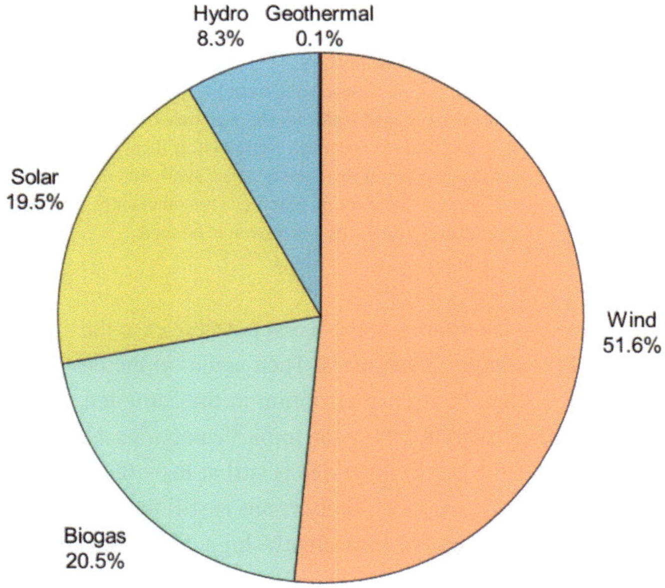

Figure 4.43: Composition of a renewable energy mix based on a real renewable energy mix by a German provider from the year 2020 [167].

When using a renewable electricity mix, the performance of the closed cooling system improves further compared to the open cooling system as shown in Figure 4.44. While the average impact reduction is 10.0 % when using the German electricity mix, the average impact reduction increases to 14.3 % when using a renewable electricity mix.

Especially in the impact category ozone depletion, the closed cooling system has a significant advantage over the open cooling system with a reduction of 26.6 % when using the renewable energy mix. Only in the category resource use of mineral and metal, the difference between both systems is negligible if both systems use renewable energy.

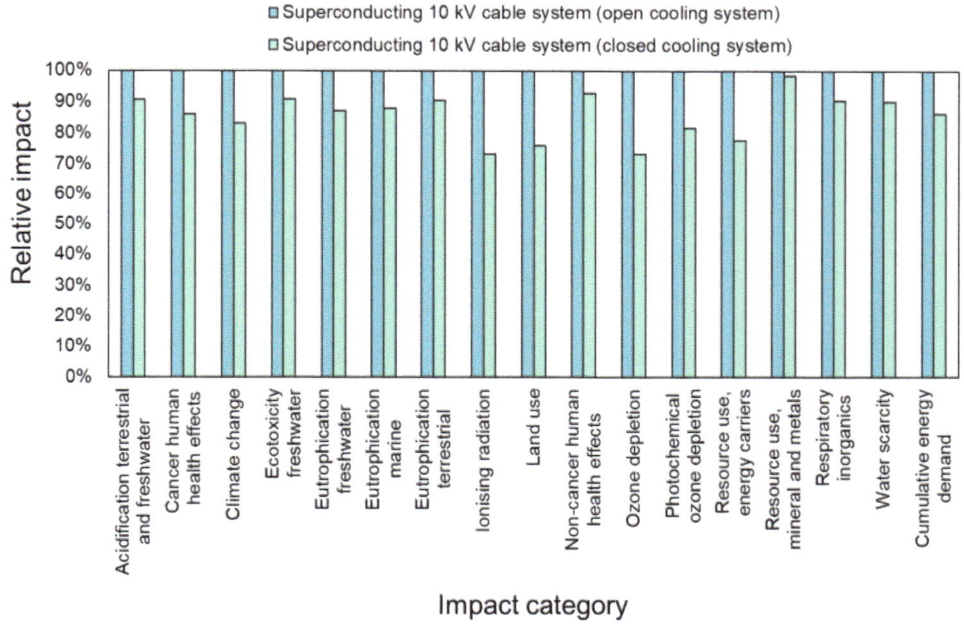

Figure 4.44: Relative impact comparison of the superconducting 10 kV cable system with an open cooling system and the superconducting 10 kV cable system with a closed cooling system for each impact category at a load factor of $m_a = 0.7$. For the open cooling system liquid nitrogen is produced using a renewable energy mix, while for the closed system the cryocooler uses the same renewable energy mix. The cable system with the higher impact in each category is indicated by the given value of 100 %. The corresponding impact of the other cable system is given as a relative share of 100 %.

While the comparison between both systems is important to decide which system to use, it is more important to compare the effects of the chosen energy mix. Figure 4.45 shows the comparison between the open and the closed cooling system when using both, the German electricity mix and the renewable electricity mix.

It is shown, that in the categories acidification terrestrial and freshwater, cancer human health effects, eutrophication terrestrial, and respiratory inorganics the use of the renewable energy mix leads to an increase in environmental impacts. In the other 13 impact categories, the renewable energy mix further decreases the environmental impacts of both cooling systems. Especially in the category resource use of mineral and metals a significant impact decrease can be achieved by using a renewable electricity mix. This is important as resource use of mineral and metals is the only category where the superconducting cable cannot reach any break-even point with the conventional 110 kV cable system, even if the best possible case is assumed. While the use of a renewable energy mix does not lead to lower impacts compared to the conventional cable system, the difference could be decreased by one order of magnitude.

Other categories, where the use of renewable energy leads to significant impact reductions are climate change, eutrophication freshwater, ionising radiation, and resource use of energy carriers.

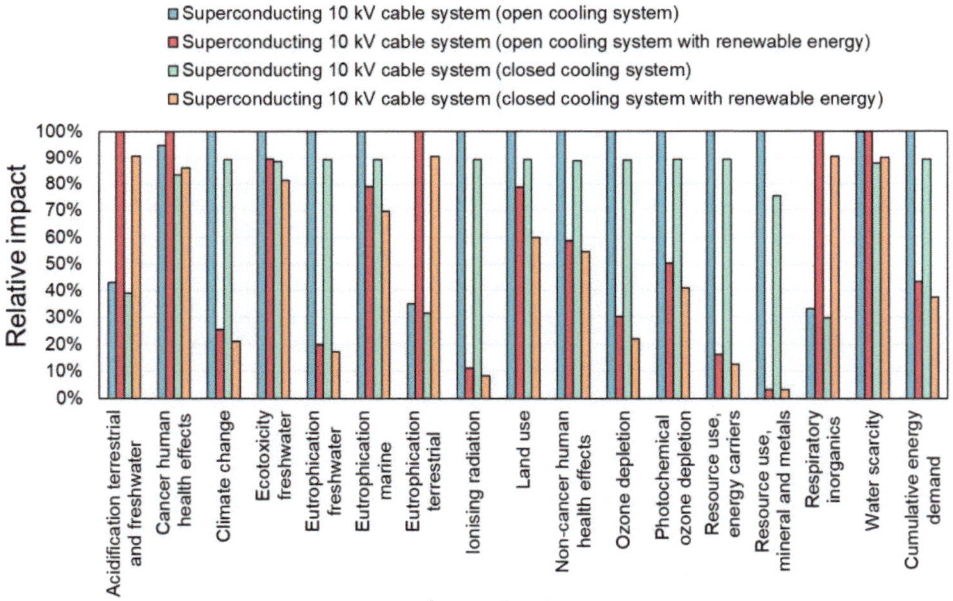

Figure 4.45: Relative impact comparison of the superconducting 10 kV cable system with an open cooling system and the superconducting 10 kV cable system with a closed cooling system for each impact category at a load factor of $m_a = 0.7$. For each cooling system there is one bar for the German electricity mix and one for the use of a renewable energy mix. The cable system with the higher impact in each category is indicated by the given value of 100 %. The corresponding impact of the other cable system is given as a relative share of 100 %.

4.6 Interpretation

In this study, a life cycle assessment was carried out for three different cable systems designed to transmit energy of up to 40 MVA over a distance of one kilometre. The three cable systems are a superconducting 10 kV cable, a conventional 110 kV cable and a cable system consisting of five conventional 10 kV cables. As a functional unit the annual transmitted electricity was chosen and in order to take load fluctuations into account the calculation was done for various load factors.

In a first comparison, each of the three cable systems was used to transmit the electricity between two substations. Each system included the same two transformers which could therefore be neglected in the comparison. First and foremoste, the use phase of each cable system was identified as the major contributor the impacts of each of the cable systems. Additionally, it was shown, that superconducting 10 kV cables are a more environmentally friendly option compared to conventional 10 kV cables. On average, the performance of the superconducting cables is 34.2 % better than the conventional medium voltage alternative at a load of $m_a = 0.7$. Therefore, if such conventional medium voltage cables were to be replaced in the future it would be a more ecological option to install superconducting 10 kV cables instead.

When compared to conventional 110 kV cables however, the superconducting cables fell short in terms of environmental impacts. On average, the environmental impacts of the superconducting cable are 788.9 % higher at a load of $m_a = 0.7$. This was due to the necessary constant cooling of the superconducting system and the higher system losses. Therefore, replacing only conventional 110 kV cables with superconducting 10 kV cables would not result in environmental advantages without adjusting the system but cause the opposite effect.

In urban networks it was shown, that with superconducting cables new network structures with less transformers are possible while keeping the n-1 redundancy. Therefore, a new configuration was investigated as well. Instead of using two transformers to go from 380 kV to 110 kV and then further down to 10 kV, one 380/10 kV transformer could be used for the superconducting medium voltage cable. In doing so, the losses of one transformer could be saved. These losses however must then be considered for the conventional 110 kV cable system. With such a transformer configuration, it was demonstrated that the superconducting cable system can be more environmentally friendly than the conventional high voltage system as long as the load is sufficiently high, meaning that m_a should at least be in the range of 0.4 to 0.6. Under the described circumstances, the superconducting cable can save about 25.3 % in environmental impacts at a load of 0.7.

Another investigation that was done, was the comparison of a closed cooling system with an open cooling system. It turned out, that the closed cooling system was the environmentally friendlier cooling system that could reduce the environmental impacts by 10 %. It was also shown that in combination with the adjusted transformer configuration the closed cooling system could increase the ecological advantage of the superconducting cable system compared to the conventional high voltage one.

In an additional scenario analysis, the use of renewable energy mixes was investigated. It was demonstrated that the renewable energy mix decreases the environmental impact in most impact categories making it the preferrable choice. Especially when using the closed cooling system where the system operator of the superconducting cable can influence the electricity mix, it proved to be the better choice to decrease environmental impacts. Compared to using an open cooling system with the German electricity mix, using a closed cooling system with a renewable energy mix reduces the impact in 14 impact categories on average by 56.5 %.

The aim of this study was to analyse if superconducting cables could be an ecological alternative to conventional copper cables and if so under which circumstances. It was shown that on the same voltage level, superconducting cables already prove to be a viable option. When compared to higher voltage cables, it was necessary to change the transformer configuration as the superconducting cable was only better under these conditions. Using a closed cooling system and renewable energy for the cryocooler, the superconducting medium voltage cable may very well provide a more ecological alternative compared to the conventional high voltage cable.

However, it must be noted that there were a few limitations to this study. First and foremost, for the production phase of the product life cycle of each cable system only the material amount was considered. This was because no cable manufacturer was willing to provide actual data in terms of energy consumption during production. Thus, this study did not include the energy that was necessary to produce and install the cables. Additionally, this did also imply that there is no material wasted during the production as only the material that was part of the final product was considered. Therefore, it can safely be assumed that the true environmental impacts of each cable system are slightly higher than shown in this study.

Furthermore, for some components it was necessary to use literature values as proxy data. However, the chosen literature values were from very similar components and as the production phase only had minor impact in most environmental impact categories, approximate values can be considered as sufficient.

Additionally, this study was based on the AmpaCity cable in Essen, Germany. This cable was planned and installed as a test cable to demonstrate the feasibility of using a superconducting cable in an inner-city electricity grid. The cable was therefore not yet a perfectly optimised cable that should be in operation for longer than the project duration. Especially the thermal losses where higher than they needed to be. In a feasibility study, it was assumed that in the future the losses of the cable could potentially be reduced by more than 60 % [162]. Therefore, it can safely be assumed that the environmental impacts of superconducting medium voltage cables will further decrease in the future. This would make superconducting medium voltage cables an even better alternative to conventional medium voltage cables. A replacement of conventional high voltage cables with superconducting medium voltage cables will not improve the environmental impact due to the higher losses at lower voltages. By using the system benefits that superconducting cables enable a further reduction of the number of transformers is viable. In this configuration superconducting cables offer a lower environmental impact even at medium load factors.

5 Summary, Conclusions and Outlook

In this study, a prospective life cycle assessment of rare earth barium copper oxide high-temperature superconductors and their future grid applications in superconducting power cables for energy distribution was conducted. Prospective life cycle assessment introduce new problems to the method of life cycle assessment. Due to assessing future developments instead of a current state of a product, a prospective modell inherits a higher uncertainty. To address this problem, multiple industry experts were contacted during this study. These experts provided data about potential or planned future developments of the production and application of high-temperature superconductors.

In the first cradle-to-gate life cycle assessment of this study, two ways to produce of second generation high-temperature superconducting tapes were examined. The first production technique was the process of inclined substrate deposition as applied by the company THEVA. This process is already in an industrial scale and the superconducting tapes are commercially available. The second examined production process was inkjet printing. This process is used by the company Oxolutia but is still on a laboratory scale.

For both tapes, the current production line was anylsed in order to identify the processes that contribute the most to the environmental impacts of the respective tapes. For the THEVA tape, the silver layer and the gadolinium barium copper oxide layer caused the highest environmental impacts with an average share of 39 % and 30 %, respectively. In case of the Oxolutia tape, the yttrium barium copper oxide layer has the highest impact with an average share of 48 % across all impact categories.

Potential future developments were analysed for both superconducting tapes in a prospective life cycle assessment. In order to estimate how the environmental impacts will change in the future, experts were contacted to assess how production will potentially develop. In case of the THEVA tape, an increase in material and energy efficiency during the production was considered. This efficiency increase leads to a decrease in greenhouse gas emissions of about 33 %. In other categories, such as the resource use of minerals and metals, the effect of the efficiency increase is negligible. For the Oxolutia tape, it was assumed that future developments would increase the production yield to 60 % while also adding a silver stabilisation layer. In total, the greenhouse gas emissions are decreased by about 74 %. However, due to the added silver layer the resource use of minerals and metal is increased significantly.

In direct comparison, the THEVA production routine currently has on average 79 % less environmental impacts in 14 impact categories. Considering potential future developments decrease the difference between both production processes slightly but results in the THEVA tape performing better in all impact categories.

Both superconductors were also compared to the production of a conventional copper conductor. The production of the superconducting tape by THEVA has less environmental impacts than the production of a copper conductor with the same current carrying capacity. However, this comparison has its limits as both conductor types, superconducting and conventional, behave differently during their use phase. The superconducting tape requires cooling, whereas the conventional conductor has ohmic losses due to its inherent resistance. Therefore, in order to compare both conductors in a fair way, the use phase must be included in the comparison.

Thus, the application of high-temperature superconductors in power cables was analysed in a second prospective life cycle assessment and compared to conventional cable alternatives. In a first comparison, the use of a 1 km long, 10 kV superconducting cable was compared with a conventional 110 kV cable and five 10 kV cables. All three systems can transmit a power of up to 40 MVA. However, the superconducting cable requires less space than the conventional cables and can transmit the electricity on medium voltage without the need for multiple cables.

For all three cable systems, the use phase is identified as the life cycle phase with the highest environmental impacts. In case of the conventional cables, this is due to the losses that occur during the use phase. Apart from grid losses, the superconducting cable system also requires constant cooling during the use phase. As a result, the use phase causes about 80 % of the annual environmental impacts of a superconducting cable system.

In direct comparison, it was shown that a superconducting medium voltage cable can be an environmentally friendlier alternative to conventional medium voltage cable. To achieve this benefit, the load factor must be sufficiently high. This load factor represents the ratio between the transmitted and the maximum electricity over time and should at least be at $m_a = 0.5$. However, when compared to a conventional high voltage cable, the superconducting cable performs worse regardless of the cable load.

Thus, a scenario analysis was conducted in which the transformer configuration of the superconducting cable was changed from two transformers to only one transformer. This scenario represents system benefits of superconducting cables that can lead to savings in transformer requirements compared to conventional cable systems. In this scenario, the use of a superconducting cable system leads to an average of 32 % less environmental impacts in 14 impact categories.

As the use phase and especially the required liquid nitrogen production were identified as the main source for environmental impacts of the superconducting cable, the potential use of an alternative cooling system was also analysed. This alternative cooling system uses electricity to re-cool the liquid nitrogen and thus removes the necessity to constantly replenish the cooling liquid. The use of such a closed cooling system further reduces the environmental impacts of the superconducting cable by about 10 %. In addition, using electricity from renewable sources to operate the cryocooler can improve these savings to more than 50 %.

Therefore, while still being an emerging technology superconducting cable systems are a promising technology for a future sustainable electricity grid. This study identified not only potential future improvements in the production of high-temperature superconducting tapes but also analysed under which circumstances their application in power cables can be an environmentally friendly alternative to conventional power cables. There are a few main goals that tape and cable manufacturers should focus on to further establish superconducting cables for inner city energy distribution:

- Firstly, increasing the material and energy efficiency during the production of superconducting tapes as well as the current-carrying capacity of the tapes. This would not only make the production more environmentally friendly but would also result in less tapes being required during the use phase. Furthermore, this would also reduce the amount of HTS tape that must be treated at the end of life of the cable. This is important as the tapes are currently melted down as steel scrap during recycling and thus valuable and critical materials are lost within that process. Thus, increasing he efficiency also reduces the material loss.
- Secondly, increasing the efficiency of the superconducting cables. Lower losses and better thermal insulation reduces the required cooling energy during the use phase, which was identified as the main contributor to the environmental impacts.
- Thirdly, using a closed cooling system that is ideally operated with electricity from renewable sources. Superconducting cables require constant cooling. Therefore, using more efficient system components can further reduce the environmental impacts of superconducting cable systems.

This study also identified potential future research fields and knowledge gaps. The data that was used in this study is partially based on literature values. For future works, the results could be enhanced. For example, this could be achieved by including on-site data from the superconducting cable production. Additionally, the production of substrates for superconducting tapes, such as Hastelloy® C-276 and sapphire, could be analysed in an in-depth life cycle assessment.

Furthermore, the AmpaCity cable that was used as a case study is a 1 km long, 10 kV superconducting cable. However, there are already projects that plan on installing longer superconducting cable systems. For example, the goal of the SuperLink project is to install a 12 km long cable in the inner city of Munich, Germany. As not only the losses of such a cable differ from the AmpaCity cable but also the cooling system has to be designed and scaled differently, future life cycle assessments could analyse the effect of such an increased cable length on the environmental impacts when compared to conventional cables.

In addition, the SuperLink cable will be a high voltage cable. This study identified that even a medium voltage superconducting cable could compete with a conventional high voltage cable given the right circumstances. Therefore, the comparison of a superconducting high voltage cable with a conventional high voltage cable could potentially identify further benefits of an application of superconductors.

Last but not least, the AmpaCity cable is an AC cable which results in higher losses compared to DC applications. Thus, the life cycle assessment of a superconducting DC cable should also be conducted in the future.

This study also demonstrated the importance of high data quality and the use of primary data from the industry to ensure low data uncertainty. Additionally, it was shown that a prospective life cycle assessment can not only identify potential environmental weak spots early on but also give recommendations on planned future technology developments. Therefore, life cycle assessments can help to further establish high-temperature superconductors as potentially environmentally-friendly technology in specific application fields. Furthermore, potential environmental hot-spots that should be improved can be identified early on.

References

[1] United Nations, "Paris Agreement," Paris, 2015.

[2] Intergovernmental Panel on Climate Change (IPCC), "Climate Change 2021: The Physical Science Basis. Contribution of Working Group I to the Sixth Assessment Report of the Invergovernmental Panel on Climate Change," Cambridge University Press. In Press., 2021.

[3] statista, "Energy-related carbon dioxide emissions worldwide from 1975 to 2021*," 2021. [Online]. Available: https://www.statista.com/statistics/526002/energy-related-carbon-dioxide-emissions-worldwide/. [Accessed September 2021].

[4] H. Ritchie and M. Roser, "CO2 and Greenhouse Gas Emissions," 2020. [Online]. Available: https://ourworldindata.org/co2-and-other-greenhouse-gas-emissions. [Accessed September 2020].

[5] German Environment Agency, "Indicator: Greenhouse gas emissions," 2021. [Online]. Available: https://www.umweltbundesamt.de/en/data/environmental-indicators/indicator-greenhouse-gas-emissions#at-a-glance. [Accessed September 2021].

[6] United States Environmental Protection Agency, "Sources of Greenhouse Gas Emissions," 2021. [Online]. Available: https://www.epa.gov/ghgemissions/sources-greenhouse-gas-emissions. [Accessed September 2021].

[7] Federal Ministry for the Environment, Nature Conservation and Nuclear Safety, "Strombedarf und Netze: Ist das Stromnetz fit für die Elektromobilität? (German)," 2020. [Online]. Available: https://www.bmu.de/themen/luft-laerm-mobilitaet/verkehr/elektromobilitaet/strombedarf-und-netze. [Accessed September 2021].

[8] P. K. Roy Chowdhury, J. E. Weaver, E. M. Weber, D. Lunga, S. T. M. LeDoux, A. N. Rose and B. L. Bhaduri, "Electricity consumption patterns within cities: Application of data-driven sttlement charaterization method," *International Journal of Digital Earth,* vol. 13, no. 1, pp. 119-135, 2020.

[9] M. Stemmle, F. Merschel, M. Noe, L. Hofmann and A. Hobl, "Superconducting MV cables to replace HV cables in urban area distribution grids," *IEEE PES Transmission and Distribution Conference and Exposition,* pp. 1-5, 2012.

[10] H. Kammerling Onnes, "Further experiments with Liquid Helium. G. On the Electrical Resistance of Pure Metals, etc. VI. On the Sudden Change in the Rate

at which the Resistance of Mercury Disappears," *Communications - Leiden 124c,* 1911.

[11] J. G. Bednorz and K. A. Müller, "Possible High Tc Superconductivity in the Ba-La-Cu-O System," *Zeitschrift für Physik B - Condensed Matter,* vol. 64, pp. 189-193, 1986.

[12] B. Keimer, S. A. Kivelson, M. R. Norman, S. Uchida and J. Zaanen, "From quantum matter to high-temperature superconductivity in copper oxides," *Nature,* vol. 518, pp. 179-186, 2015.

[13] S. Nishijima, S. Eckroad, A. Marian, K. Choi, W. S. Kim, M. Terai, Z. Deng, J. Zheng, J. Wang, K. Umemoto, J. Du, P. Febvre, S. Keenan, O. Mukhanov, L. D. Cooley, C. P. Foley, W. V. Hassenzahl and M. Izumi, "Superconductivity and the environment: a Roadmap," *Superconductor Science and Technology,* vol. 26, pp. 113001: 1-35, 2013.

[14] P. D. Baumann, "Energy conservation and environmental benefits that may be realized from superconducting magnetic energy storage," *IEEE Transactions on Energy Conversion,* vol. 7, no. 2, pp. 253-259, 1992.

[15] R. A. Hawsey and S. Morozumi, "The energy and environmental benefits of superconducting power products," *Mitigation and Adaptation Strategies for Global Change,* vol. 10, pp. 279-306, 2005.

[16] T. Hartikainen, R. Mikkonen and J. Lehtonen, "Environmental advantages of superconducting devices in distributed electricity-generation," *Applied Energy,* vol. 84, pp. 29-38, 2007.

[17] S. Kamiya, E. Tanaka, H. Kita, J. Hasegawa and T. Shirasaki, "Life Cycle Assessment and Economical Evaluation of Superconducting Magnetic Energy Storage Systems in a Power System," *Electrical Engineering in Japan,* vol. 131, no. 1, pp. 1302-1311, 2000.

[18] J. Lloberas-Valls, G. B. Pérez and O. Gomis-Bellmunt, "Life-Cycle Assessment Comparison Between 15-MW Second-Generation High-Temperature Superconductor and Permanent-Magnet Direct-Drive Synchonous Generators for Offshore Wind Energy Applications," *IEEE Transactions on Applied Superconductivity,* vol. 25, no. 6, pp. 5204209: 1-9, 2015.

[19] R. Berti, F. Barberis, V. Rossi and L. Martini, "Comparison of the ecoprofiles of superconducting and conventional 25 MVA transformers using the life cycle assessment methodology," *CIRED 2009 - 20th International Conference on Electricity Distribution,* p. Paper 0773, 2009.

[20] A. Marian, A. Ballarino, C. Catalan, N. Dittmar, G. Escamez, S. Gianelli, F. Grilli, S. Holé, C. Haberstroh, F. Lesur, C. Poumarède, M. Tropeano, G. Vega, Bruzek. and C.-E., "An MgB2 HVDC Superconducting Cable for Power Transmission with a Reduced Carbon Footprint," in *Eco-design in Electrical Engineering,*

Springer International Publishing, 2018, pp. 129-135.

[21] R. G. Hunt and W. E. Franklin, "LCA - How it Came About - Personal Reflections on the Origin and the Development of LCA in the USA," *International Journal of Life Cycle Assessment,* vol. 1, no. 1, pp. 4-7, 1996.

[22] W. Klöpffer, "Life Cycle Assessment - From the Beginning to the Current State," *Environmental Science and Pollution Research,* vol. 4, no. 4, pp. 223-228, 1997.

[23] J. B. Guinée, R. Heijungs, G. Huppes, A. Zamagni, P. Masoni, R. Buonamici, T. Ekvall and T. Rydberg, "Life Cycle Assessment: Past, Present and Future," *Environmental Science & Technology,* vol. 45, no. 1, pp. 90-96, 2011.

[24] Life Cycle Initiative, "What is Life Cycle Thinking?," [Online]. Available: https://www.lifecycleinitiative.org/starting-life-cycle-thinking/what-is-life-cycle-thinking/. [Accessed April 2021].

[25] D. Postlethwaite, "Development of Life Cycle Assessment (LCA) - The Role of SETAC (Society of Environmental Toxicology and Chemistry) and the 'Code of Practice'," *Environmental Science and Pollution Research,* vol. 1, no. 1, pp. 54-55, 1994.

[26] SETAC, "Guidelines for Life-Cycle Assessment: A 'Code of Practice' From the workshop held at Sesimbra, Portugal, 31 March - 3 April 1993," *Environmental Science and Pollution Research,* vol. 1, p. 55, 1994.

[27] DIN Deutsches Institut für Normung e.V., Ed., DIN EN ISO 14040 Environmental management - Life cycle assessment - Principles and framework (ISO 14040:2006), Berlin: Beuth-Verlag, 2009.

[28] DIN Deutsches Institut für Normung e.V., Ed., DIN EN ISO 14044 Environmental management - Life Cycle Assessment - Requirements and guidelines (ISO 14044:2006), Berlin: Beuth-Verlag, 2006.

[29] M.-A. Wolf, R. Pant, K. Chomkhamsri, S. Sala and D. Pennington, "JRC Reference Reports - The International Reference Life Cycle Data System (ILCD) Handbook - Towards more sustainable production and consumption for a resource-efficient Europe," EUR 24982 EN. Publications Office of the European Union., Luxembourg, 2012.

[30] European Commission - Joint Research Centre - Institute for Environment and Sustainability, "International Reference Life Cycle Data System (ILCD) Handbook - General guide for Life Cycle Assessment - Detailed guidance," First edition March 2010. EUR 24708 EN. Publications Office of the European Union, Luxembourg, 2010.

[31] European Commission - Joint Research Centre - Institute for Environment and Sustainability, "International Reference Life Cycle Data System (ILCD) Handbook - Recommendations for Life Cycle Impact Assessment in the

European Context," First edition November 2011. EUR 24571 EN. Publications Office of the European Union, Luxembourg, 2011.

[32] European Commission - Joint Research Centre - Institute for Environment and Sustainability, "International Reference Life Cycle Data System (ILCD) Handbook - Specific guide for Life Cycle Inventory data sets," First edition March 2010. EUR 24709 EN. Publications Office of the European Union, Luxembourg, 2010.

[33] European Commission - Joint Research Centre - Institute for Environment and Sustainability, "International Reference Life Cycle Data System (ILCD) Handbook - Framework and requirements for Life Cycle Impact Assessment models and indicators.," First Edition March 2010. EUR 24586 EN. Publications Office of the European Union., Luxembourg, 2010.

[34] A.-M. Tillman, M. Svingby and H. Lundström, "Life Cycle Assessment of Municipla Waste Water Systems," *International Journal of Life Cycle Assessment,* vol. 3, no. 3, pp. 145-157, 1998.

[35] T. Li, H. Zhang, Z. Liu and Q. Ke, "A system boundary identification method for life cycle assessment," *International Journal of Life Cycle Assessment,* vol. 19, no. 3, pp. 646-600, 2014.

[36] M. A. Curran, "Chapter 1 Overview of Goal and Scope Definition in Life Cycle Assessment," in *Goal and Scope Definition in Life Cycle Assessment*, M. A. Curran, Ed., Springer Nature, 2017, pp. 1-62.

[37] A. Bjørn and M. Z. Hauschild, "Cradle to Cradle and LCA," in *Life Cycle Assessment - Theory and Practice*, M. Z. Hauschild, R. K. Rosenbaum and S. I. Olsen, Eds., Switzerland, Springer International Publishing, 2018, pp. 605-631.

[38] E. C. Peereboom, R. Kleijn, S. Lemkowitz and S. Lundie, "Influence of Inventory Data Sets on Life-Cycle Assessment Results: A Case Study," *Journal of Industrial Ecology,* vol. 2, no. 3, pp. 109-130, 1999.

[39] N. W. Ayer, P. H. Tyedmers, N. L. Pelletier, U. Sonesson and A. Scholz, "Co-Product Allocation in Life Cycle Assessments of Seafood Production Systems: Review of Problems and Strategies," *International Journal of Life Cycle Assessment,* vol. 7, pp. 480-487, 2007.

[40] A. S. Kaufman, P. J. Meier, J. C. Sinistore and D. J. Reinemann, "Applying life-cycle assessment to low carbon fuel standards - How allocation choices influence carbon intensity for renewable transportation fuels," *Energy Policy,* vol. 38, no. 9, pp. 5229-5241, 2010.

[41] E. Svanes, M. Vold and O. J. Hanssen, "Effect of different allocation methods on LCA results of products from wild-caught fish and on the use of such results," *International Journal of Life Cycle Assessment,* vol. 16, no. 6, pp. 512-521, 2011.

[42] J. C. Bare, P. Hofstetter, D. W. Pennington and H. A. Udo de Haes, "Midpoints versus Endpoints: The Sacrifices and Benefits," *International Journal of Life Cycle Assessment,* vol. 5, no. 6, pp. 319-326, 2000.

[43] Y. H. Dong and S. T. Ng, "Comparing the midpoint and endpoint approaches based on ReCiPe - a study of commercial buildings in Hong Kong," *International Journal of LIfe Cycle Assessment,* vol. 19, no. 7, pp. 1409-1423, 2014.

[44] T. Kägi, F. Dinkel, R. Frischknech, S. Humbert, J. Lindberg, S. De Mester, T. Ponsioen, S. Sala and U. W. Schenker, "Session "Midpoint, endpoint or single score for decision-making?" - SETAC Europe 25th Annual Meeting, May 5th, 2015," *International Journal of Life Cycle Assessment,* vol. 21, no. 1, pp. 129-132, 2016.

[45] M. A. J. Huijbregts, Z. J. N. Steinmann, P. M. F. Elshout, G. Stam, F. Verones, M. D. M. Vieira, A. Hollander, M. Zijp and R. van Zelm, "ReCiPe 2016 v1.1 A harmonized life cycle impact assessment method at midpoint and endpoint level. Report I: Characterization," RIVM National Institute for Public Health and the Environment, Netherlands, 2017.

[46] M. A. J. Huijbregts, Z. J. N. Steinmann, P. M. F. Elshout, G. Stam, F. Verones, M. D. M. Vieira, A. Hollander, M. Zijp and R. van Zelm, "ReCiPe2016: a harmonised life cycle impact assessment method," *International Journal of Life Cycle Assessment,* vol. 22, no. 2, pp. 138-147, 2017.

[47] S. Fazio, F. Biganzioli, V. De Laurentiis, L. Zampori, S. Sala and E. Diaconu, "Supporting information to the characterisation factors of recommended EF Life Cycle Impact Assessment methods, version 2, from ILCD to EF 3.0," EUR 29600 EN, European Commission, Ispra, 2018.

[48] R. Frischknecht, F. Wyss, S. Büsser Knöpfel, T. Lützkendorf and M. Balouktsi, "Cumulative energy demand in LCA: the energy harvested approach," *International Journal of LIfe Cycle Assessment,* vol. 20, no. 7, pp. 957-969, 2015.

[49] A. Edelen and W. Ingwersen, "Guidance on Data Quality Assessment for Life Cycle Inventory Data - Version 1," Cincinnati, Ohio, USA, 2016.

[50] G. Wernet, C. Bauer, B. Steubing, J. Reinhard, E. Moreno-Ruiz and B. Weidema, "The ecoinvent database version 3 (part I): overview and methodology," *International Journal of Life Cycle Assessment,* vol. 21, no. 9, pp. 1218-1230, 2016.

[51] Intergovernmental Panel on Climate Change (IPCC), Climate Change 2013 The Physical Science Basis - Working Group I Contribution to the Fifth Assessment Report of the Intergovernmental Panel on Climate Change, T. F. Stocker, D. Qin, G. Plattner, M. M. Tignor, S. K. Allen, J. Boschung, A. Nauels, Y. Xia, V. Bex and P. M. Midgley, Eds., Cambridge, United Kingdom and New York, NY, USA: Cambridge University Press, 2013.

[52] J. F. Dean, J. J. Middelburg, T. Röckmann, R. Aerts, L. G. Blauw, M. Egger, M. S. M. Jetten, A. E. E. de Jong, O. H. Meisel, O. Rasigraf, C. P. Slomp, M. H. in't Zandt and A. J. Dolman, "Methane Feedbacks to the Global Climate System in a Warmer World," *Review of Geophysics,* vol. 56, no. 1, pp. 207-250, 2018.

[53] T. E. Graedel, J. Allwood, J.-P. Birat, M. Buchert, C. Hagelüken, B. K. Reck, S. F. Sibley and G. Sonnemann, "Recycling Rates of Metals - A Status Report," United Nations Environment Programme, International Resource Panel, Working Group on the Global Metal Flows, 2011.

[54] European Commission - Joint Research Centre - Institute for Environment and Sustainability, "The International Reference Life Cycle Data System (ILCD) Handbook - Towards more sustainable production and consumption for a resource-efficient Europe," First edition 2012. EUR 24982 EN. Publications Office of the European Union, Luxembourg, 2012.

[55] R. K. Rosenbaum, M. Hauschild, A.-M. Boulay, P. Fantke, A. Laurent, M. Núñez and M. Vieira, "Chapter 10 Life Cycle Impact Assessment," in *Life Cycle Assessment,* M. Z. Hauschild, S. I. Olsen and R. K. Rosenbaum, Eds., Switzerland, Springer Nature, 2018, pp. 167-270.

[56] R. Frischknecht, O. Jolliet, L. Milà i Canals, S. Pfister, A. Sahnoune, C. Ugaya and B. Vigon, "Motivation, Context and Overview," in *Global Guidance for Life Cycle Impact Assessment Indicators Volume 1,* United Nations Environment Programme - Society of Environmental Toxicology and Chemistry - Life Cycle Initiative, 2018, pp. 31-39.

[57] L. Tikana, H. Sievers and A. Klassert, "Life Cycle Assessment of Copper Products," Deutsches Kupferinstitut - Life Cycle Centre, Düsseldorf, Germany, 2005.

[58] European Commission, "Communication from the Commission to the European Parliament and the Council - Building the Single Market for Green Products - Facilitating better information on the environmental performance of products and organisations," COM(2013) 196 final, Brussels, 2013.

[59] J. Seppälä, M. Posch, M. Johansson and J.-P. Hettelingh, "Country-Dependent Characterisation Factors for Acidification and Terrestrial Eutrophication Based on Accumulated Exceedance as an Impact Category Indicator," *International Journal of Life Cycle Assessment,* vol. 11, no. 6, pp. 403-416, 2006.

[60] M. Posch, J. Seppälä, J.-P. Hettelingh, M. Johansson, M. Margni and O. Jolliet, "The role of atmospheric dispersion models and ecosystem sensitivity in the determination of characterisation factors for acidifying and eutrophying emissions in LCIA," *International Journal of LIfe Cycle Assessment,* vol. 13, no. 6, pp. 477-486, 2008.

[61] R. K. Rosenbaum, T. M. Bachmann, L. Swirsky Gold, M. A. J. Huijbregts, O. Jolliet, R. Juraske, A. Koehler, H. F. Larsen, M. MacLeod, M. Margni, T. E. McKone, J. Payet, M. Schuhmacher, D. van de Meent and M. Z. Hauschild, "USEtox - the

UNEP-SETAC toxicity model: recommended characterisation factors for human toxicity and freshwater ecotoxicity in life cycle impact assessment," *International Journal of Life Cycle Assessment,* vol. 13, no. 7, pp. 532-546, 2008.

[62] J. Struijs, A. van Dijk, H. Slaper, H. J. van Wijnen, G. J. M. Velders, G. Chaplin and M. A. J. Huijbregts, "Spatial- and Time-Explicit Human Damage Modeling of Ozone Depleting Substances in Life Cycle Impact Assessment," *Environmental Science & Technology,* vol. 1, pp. 204-209, 2010.

[63] R. Frischknecht, A. Braunschweig, P. Hofstetter and P. Suter, "Human health damages due to ionising radiation in life cycle impact assessment," *Environmental Impact Assessment Review,* vol. 20, no. 2, pp. 159-189, 2000.

[64] U. Bos, R. Horn, T. Beck, J. P. Lindner and M. Fischer, LANCA. Characterization Factors for Life cycle Impact Assessment, Version 2.0, F. IBP, Ed., Stuttgart: Fraunhofer Verlag, 2016.

[65] V. De Laurentiis, M. Secchi, U. Bos, R. Horn and A. Laurent, "Soil quality index: Exploring options for a comprehensive assessment of land use impacts in LCA," *Journal of Cleaner Production,* vol. 215, pp. 63-74, 2019.

[66] World Meteorological Organisation (WMO), "Scientific Assessment of Ozone Depletion: 2014, World Meterological Organization, Global Ozone Research and Monitoring Project - Report Nr. 55," Geneva, Switzerland, 2014.

[67] M. Goedkoop, R. Heijungs, M. Huijbregts, A. De Schryver, J. Struijs and R. van Zelm, "ReCiPe 2008 A life cycle impact assessment method which comprises harmonised category indicators at the midpoint and the endpoint level, First edition, Report I: Characterisation," 2009.

[68] R. van Zelm, M. A. J. Huijbregts, H. A. den Hollander, H. A. van Jaarsveld, F. J. Sauter, J. Struijs, H. J. van Wijnen and D. van de Meent, "European characterizationf actors for human health damage of PM10 and ozone in life cycle impact assessment," *Atmospheric Environment,* vol. 42, no. 3, pp. 441-453, 2008.

[69] L. van Oers, A. de Koning, J. B. Guinée and G. Huppes, "Abiotic resource depletion in LCA - Improving characterisation factors for abiotic resource depletion as recommended in the new Dutch LCA Handbook," Road and Hydraulic Engineering Institute, 2002.

[70] P. Fantke, J. Evans, N. Hodas, J. Apte, M. Jantunen, O. Jolliet and T. McKone, "Health impacts of fine particulate matter," in *Global Guidance for Life Cycle Impact Assessment Indicators*, vol. 1, R. Frischknecht and O. Jolliet, Eds., United Nations Environment Programme (UNEP) / Society of Environmental Toxicology and Chemistry (SETAC) Life Cycle Initiative, 2016, pp. 76-99.

[71] A.-M. Boulay, J. Bare, L. Benini, M. Berger, M. J. Lathuillière, A. Manzardo, M.

Margni, M. Motoshita, M. Núñez, A. V. Pastor, B. Ridoutt, T. Oki, S. Worbe and S. Pfister, "The WULCA consensus characterization model for water scarcity footprints: assessing impacts of water consumption based on available water remaining (AWARE)," *International Journal of Life Cycle Assessment,* vol. 23, no. 2, pp. 368-378, 2018.

[72] S. Raychaudhuri, "Introduction to Monte Carlo simulation," *Proceedings of the 2008 Winter Simulation Conference,* pp. 91-100, 2008.

[73] M. Raynolds, M. D. Checkel and R. A. Fraser, "Application of Monte Carlo Analysis to Life Cycle Assessment," *SAE Technical Paper Series,* pp. 1-9, 1999.

[74] Y. Qin and S. Suh, "What distribution function do life cycle inventories follow?," *International Journal of Life Cycle Assessment,* vol. 22, no. 7, pp. 1138-1145, 2017.

[75] S. Muller, C. Mutel, P. Lesage and R. Samson, "Effects of Distribution Choice on the Modeling of Life Cycle Inventory Uncertainty," *Journal of Industrial Ecology,* vol. 22, no. 2, pp. 300-313, 2017.

[76] M. A. Raynolds, M. D. Checkel and R. A. Fraser, "Uncertainty, Sensitivity and Data Quality Assessment for Life Cycle Value Assessment (LCVA)," *SAE Technical Paper Series,* pp. 1-8, 1998.

[77] A. E. Björklund, "Survey of Approaches to Improve Reliability in LCA," *International Journal of Life Cycle Assessment,* vol. 7, no. 2, pp. 64-72, 2002.

[78] N. Thonemann, A. Schulte and D. Maga, "How to Conduct Prospective Life Cycle Assessment for Emerging Technologies? A Systematic Review and Methodological Guidance," *sustainability,* vol. 12, no. 3, pp. 1192, 1-23, 2020.

[79] S. Peng, T. Li, Y. Wang, Z. Liu, G. Z. Tan and H. Zhang, "Propsective Life Cycle Assessment Based on System Dynamics Approach: A Case Study on the Large-Scale Centrifugal Compressor," *Journal of Manufacturing Science and Engineering,* vol. 141, no. 2, p. 021003, 2019.

[80] D. García-Gusano, "Prospective life cycle assessment of the Spanish electricity production," *Renewable and Sustainable Energy Reviews,* vol. 75, pp. 21-34, 2017.

[81] R. Arvidsson, A.-M. Tillman, B. A. Sandén, M. Janssen, A. Nordelöf, D. Kushnir and S. Molander, "Environmental Assessment of Emerging Technologies," *Journal of Industrial Ecology,* vol. 22, no. 6, pp. 1286-1294, 2017.

[82] SimaPro, "LCA software to help you drive change," 2021. [Online]. Available: https://simapro.com. [Accessed September 2021].

[83] sphera, "GaBi Software," 2021. [Online]. Available: https://gabi.sphera.com/international/software/gabi-software/. [Accessed

September 2021].

[84] iPoint-systems, "Umberto LCA+," 2021. [Online]. Available: https://www.ifu.com/umberto/lca-software/. [Accessed September 2021].

[85] "openLCA," Greendelta, [Online]. Available: https://www.openlca.org/. [Accessed Feb 2021].

[86] "ecoinvent," ecoinvent, [Online]. Available: https://www.ecoinvent.org/. [Accessed Feb 2021].

[87] "Allocation cut-off by classification," ecoinvent, [Online]. Available: https://www.ecoinvent.org/database/system-models-in-ecoinvent-3/cut-off-system-model/allocation-cut-off-by-classification.html. [Accessed Feb 2021].

[88] V. E. Sytnikov, V. S. Vysotsky, I. P. Radchenko and N. V. Polyakova, "1G versus 2G - comparison from the practical standpoint for HTS power cable use," *Journal of Physics: Conference Series,* vol. 97, pp. 012058: 1-4, 2008.

[89] A. P. Malozemoff, S. Fleshler, M. Rupich, C. Thieme, X. Li, W. Zhang, A. Otto, J. Maguire, D. Folts, J. Yuan, H.-P. Kraemer, W. Schmidt, M. Wohlfart and H.-W. Neumueller, "Progress in high temperature superconductor coated conductors and their applications," *Superconductor Science and Technology,* vol. 21, no. 3, pp. 034005: 1-7, 2008.

[90] M. Dürrschnabel, Z. Aabdin, M. Bauer, R. Semerad, W. Prusseit and O. Eibl, "DyBa2Cu3O7-x superconducting coated conductors with critical currents exceeding 1000 A/cm," *Superconductor Science and Technology,* vol. 25, no. 10, pp. 105007: 1-4, 2012.

[91] M. Rupich, W. Zhang, X. Li, T. Kodenkandath, D. Verebelyi, U. Schoop, C. Thieme, M. Teplitsky, J. Lynch, N. Nguyen, E. Siegal, J. Scudiere, V. Maroni, K. Venkataraman, D. Miller and T. Holesinger, "Progress on MOD/RABiTS(TM) 2G HTS wire," *Physica C,* pp. 877-884, 2004.

[92] A. Goyal, D. Norton, D. Kroeger, D. Christen, M. Paranthaman, E. Specht, J. Budai, Q. He, B. Saffian, F. List, D. Lee, E. Hatfield, P. Martin, C. Klabunde, J. Mathis and C. Park, "Conductors with controlled grain boundaries: An approach to the next generation, high temperature superconducting wire," *Journal of Materials Research,* vol. 11, pp. 2924-2940, 1997.

[93] X. Wu, S. Foltyn, P. Arendt, J. Townsend, C. Adams, I. Campbell, P. Tiwari, Y. Coulter and D. Peterson, "High current YBa2Cu3O7-x thick films on flexible nickel substrates with textured buffer layers," *Applied Physics Letters,* vol. 65, no. 15, pp. 1961-1963, 1994.

[94] S. Hopkins, D. Joseph, T. Mitchell-Williams, A. Calleja, V. Vlad, M. Vilardell, S. Ricart, X. Granados, T. Puig, X. Obradors, A. Usoskin, M. Falter, M. Bäcker and B. Glowacki, "Inkjet printing of multifiliamentary YBCO for low AC loss

coated conductors," *Journal of Physics: Conference Series,* vol. 507, no. 2, pp. 022010: 1-4, 2014.

[95] K. Reichelt and X. Jiang, "The Preparation of Thin Films by Physical Vapour Deposition Methods," *Thin Solid Films,* vol. 191, no. 1, pp. 91-126, 1990.

[96] T. Holesinger, B. Maiorov, O. Ugurlu, L. Civale, Y. Chen, X. Xiong, Y. Xie and V. Selvamanickam, "Microstructural and superconducting properties of high current metal-organic chemcial vapor deposition YBa2Cu3O7-x coated conductor wires," *Superconductor Science and Technology,* vol. 22, no. 4, pp. 045025: 1-14, 2009.

[97] X. Obradors, T. Puig, A. Pomar, F. Sandiumenge, S. Pinol, N. Mestres, O. Castano, M. Coll, A. Cavallaro, A. Palau, J. Gázquez, J. González, J. Gutiérrez, N. Romà, S. Ricart, J. Moretó, M. Rossel and G. van Tendeloo, "Chemical solution deposition: a path towards low cost coated conductors," *Superconductor Science and Technology,* vol. 17, no. 8, pp. 1055-1064, 2004.

[98] K. Leonard, S. Kang, A. Goyal, K. Yarborough and D. Kroeger, "Microstructural characterization of thick YBa2Cu3Oy-x films on improved rolling-assisted biaxially textured substrates," *Journal of Materials Research,* vol. 18, no. 7, pp. 1723-1732, 2003.

[99] G. Daly, J. Pond, M. Osofsky, J. Horwitz, R. Soulen and D. Chrisey, "Pulsed Laser Deposition of Patterned Multilayers for HTS Device Fabrication," *IEEE Transactions on Applied Superconductivity,* vol. 7, no. 2, pp. 2153-2156, 1997.

[100] Y. Iijima and K. Matsumoto, "High-temperature-superconductor coated conductors: technical progress in Japan," *Superconductor Science and Technology,* vol. 13, no. 1, pp. 68-81, 2000.

[101] M. Bauer, "Supraleiter für die Energietechnik - Überblick Technologie und Stand der Technik, ivSupra ZIEHL VII," March 2020. [Online]. Available: https://ivsupra.de/vii-ziehl-vortraege/. [Accessed May 2021].

[102] R. J. Martín-Palma and A. Lakhtakia, "Chapter 15 - Vapor-Deposition Techniques," in *Engineered Biomimicry*, Elsevier, 2013, pp. 383-398.

[103] A. Guglya and E. Lyubchenko, "Chapter 4 - Ion-beam-assisted deposition of thin films," in *Emerging Applications of Nanoparticles and Architecture Nanostructures*, Elsevier, 2018, pp. 95-119.

[104] Haynes International, "Hastelloy (R) C-276 alloy: Principal Features," 2020. [Online]. Available: https://www.haynesintl.com/docs/default-source/pdfs/new-alloy-brochures/corrosion-resistant-alloys/brochures/c-276.pdf?sfvrsn=ee7929d4_20. [Accessed April 2021].

[105] THEVA, "THEVA Pro-Line HTS Wire - General Properties," March 2020. [Online]. Available: https://www.theva.de/wp-

content/uploads/2020/02/190425_THEVA_General-Properties_Update_2020_klein.pdf. [Accessed April 2021].

[106] K. Fujino, K. Ohmatsu, Y. Sato, S. Honjo and Y. Takahashi, "Chapter 3 - Inclined Substrate Deposition," in *Second-Generation HTS Conductors*, Norwell, Massachusetts, USA, Kluwer Academic Publishers, 2005, pp. 47-52.

[107] S. Hopkins, T. Mitchell-Williams, D. Vanden Bussche, A. Calleja, V. Vlad, M. Vilardell, X. Granados, T. Puig, X. Obradors, A. Usoskin, M. Soloviov, M. Vojenciak, F. Gömöry, I. Van Driessche, M. Bächer and B. Glownacki, "Low AC Loss Inkjet-Printed Multifilamentary YBCO Coated Conductors," *IEEE Transactions on Aplied Superconductivity,* vol. 26, no. 3, pp. 6602905: 1-5, 2016.

[108] C. Lacroix, Y. Lapierre, J. Coulombe and F. Sirois, "High normal zone propagation velocity in second generation high-temperature superconductor coated conductors with a current flow diverter architecture," *Superconductor Science and Technology,* vol. 27, no. 5, pp. 055013: 1-6, 2014.

[109] H. Rijckaert, P. Cayado, R. Nast, J. Diez Sierra, M. Erbe, P. López Dominguez, J. Hänisch, K. De Buysser, B. Holzapfel and I. Van Driessche, "Superconducting HfO2-YBa2Cu3O7-x Nanocomposite Films Deposited Using Ink-Jet Printing of Colloidal Solutions," *Coatings,* vol. 10, no. 1, pp. 17: 1-15, 2020.

[110] T. Koenen, A. Undisz, S. Otten and M. Rettenmayr, "Stainless Steel as Substrate Material for High-Temperature Superconducting Tapes Process Via the ISD MgO Route," *IEEE Transactions on Applied Superconductivity,* vol. 29, no. 4, p. 6900309, 2019.

[111] G. Genchi, A. Carocci, G. Lauria, M. S. Sinicropi and A. Catalano, "Nickel: Human Health and Environmental Toxicology," *International Journal of Environmental Research and Public Health,* vol. 17, no. 3, pp. 679: 1-21, 2020.

[112] M. Mistry, J. Gediga and S. Boonzaier, "Life cycle assessment of nickel products," *International Journal of Life Cycle Assessment,* vol. 21, no. 11, pp. 1559-1572, 2016.

[113] G. Deutscher, "High-Voltage Superconducting Fault Current Limiters Based on High-Diffusivity Dielectric Substrates," *Journal of Superconductivity and Novel Magnetism,* vol. 31, no. 7, pp. 1961-1963, 2018.

[114] L. Antognazza, M. Therasse, M. Decroux, F. Roy, B. Dutoit, M. Abplanalp and Ø. Fischer, "Comparison Between the Behaviour of HTS Thin Film Grown on Sapphire and Coated Conductors for Fault Current Limiter Applications," *IEEE Transactions on Applied Superconductivity,* vol. 19, no. 3, pp. 1960-1963, 2009.

[115] T. Koenen, A. Undisz, S. Otten and M. Rettenmayr, "Stainless Steel as Substrate Material for High-Temperature Superconducting Tapes Processed Via the ISD MgO Route," *IEEE Transactions on Applied Superconductivity,* vol. 29, no. 4, p.

6900309, 2019.

[116] K. Hillier, "Magnesium Oxide," in *xPharm: The Comprehensive Pharmacology Reference*, Elsevier, 2008, pp. 1-4.

[117] J. Nobre, A. Hawreen, M. Bravo, L. Evangelista and J. de Brito, "Magnesia (MgO) Production and Characterization, and Its Influence on the Performance of Cementitious Materials: A Review," *materials,* vol. 13, no. 21, pp. 4752: 1-31, 2020.

[118] U.S. Geological Survey, "Mineral Commodity Summaries 2021," U.S. Geological Survey, Reston, Virginia, 2021.

[119] European Commission, "Study on the EU's list of Critical Raw Materials (2020) Final Report," Luxembourg: Publications Office of the European Union, 2020.

[120] B. Ma, M. Li, Y. Jee, R. Koritala, B. Fisher and U. Balachandran, "Inclined-substrate deposition of biaxially textured magnesium oxide thin films for YBCO coated conductors," *Physica C,* vol. 366, no. 4, pp. 270-276, 2002.

[121] International Union of Pure and Applied Chemistry, "Nomenclature of Inorganic Chemistry - IUPAC Recommendations 2005," Royal Society Chemistry Publishing, Cambridge, UK, 2005.

[122] J. H. L. Voncken, "Chapter 3: Physical and Chemical Properties of the Rare Earths," in *The Rare Earth Elements - An Introduction*, Switzerland, Springer International Publishing AG, 2016, pp. 53-73.

[123] P. S. Arshi, E. Vahidi and F. Zhao, "Behind the Scenes of Clean Energy: The Environmental Footprint of Rare Earth Products," *ACS Sustainable Chemistry & Engineering,* vol. 3, pp. 3311-3320, 2018.

[124] K. Zhang, A. N. Kleit and A. Nieto, "An economics strategy for criticality – Application to rare earth element Yttrium in new lighting technology and its sustainable availability," *Renewable and Sustainable Energy Reviews,* vol. 77, pp. 899-915, 2017.

[125] J. H. L. Voncken, "Chapter 5: Applications of the Rare Earths," in *The Rare Earth Elements - An Introduction*, Switzerland, Springer International Publishing AG, 2016, pp. 89-107.

[126] L. Wang, X. Huang, Y. Yu, L. Zhao, C. Wang, Z. Feng, D. Cui and Z. Long, "Towarss cleaner production of rare earth elements from bastnaesite in China," *Journal of Cleaner Production,* vol. 165, pp. 231-242, 2017.

[127] J. H. L. Voncken, "Chapter 1: The Rare Earth Elements - A Special Group of Metals," in *The Rare Earth Elements - An Introduction*, Switzerland, Springer International Publishing AG , 2016, pp. 1-15.

[128] C. Tunsu, M. Petranikova, M. Gergorić, C. Ekberg and T. Retegan, "Reclaiming rare earth elements from end-of-life products: A review of the perspectives for urban mining using hydrometallurgical unit operations," *Hydrometallurgy,* pp. 239-258, 2015.

[129] G. B. Haxel, J. B. Hedrick and G. J. Orris, "Rare Earth Elements - Critical Resources for High Technology, USGS Fact Sheet 087-02," U.S. Geological Survey, 2005.

[130] V. Balaram, "Rare earth elements: A review of applications, occurrence, exploration, analysis, recycling, and environmental impact," *Geoscience Frontiers,* vol. 10, no. 4, pp. 1285-1303, 2019.

[131] European Commission, "COMMUNICATION FROM THE COMMISSION TO THE EUROPEAN PARLIAMENT, THE COUNCIL, THE EUROPEAN ECONOMIC AND SOCIAL COMMITTEE AND THE COMMITTEE OF THE REGIONS Critical Raw Materials Resilience: Charting a Path towards greater Security and Sustainability," European Commission, COM(2020) 474 final, Brussels, 2020.

[132] G. G. Zaimes, B. J. Hubler, S. Wang and V. Khanna, "Environmental Life Cycle Perspective on Rare Earth Oxide Production," *ACS Sustainable Chemistry & Engineering,* vol. 3, no. 2, pp. 237-244, 2015.

[133] H. Sverdrup, D. Koca and K. Ragnarsdottir, "Investigating the sustainability of the global silver supply, reserves, stocks in society and market price using different approaches," *Resources, Conservationd and Recycling,* vol. 83, pp. 121-140, 2014.

[134] R. Bhattacharya, Y. Qiao and V. Selvamanickam, "Electrodeposited Cu-Stabilization Layer for High-Temperature Superconducting Coated Conductors," *Journal of Superconductivity and Novel Magnetism,* vol. 24, no. 1-2, pp. 1021-1026, 2011.

[135] N. Nassar, R. Barr, M. Browning, Z. Diao, E. Friedlander, E. Harper, C. Henly, G. Kavlak, S. Kwatra, C. Jun, S. Warren, M.-Y. Yang and T. Graedel, "Criticality of the Geological Copper Family," *Environmental Science & Technology,* vol. 46, no. 2, pp. 1071-1078, 2011.

[136] W. T. B. de Sousa, A. Kudymow, S. Strauss, S. Palasz, S. Elschner and M. Noe, "Simulation of the Thermal Performance of HTS Coated Conductors for HVDC SFCL," *Journal of Physics: Conference Series,* Vols. 1559 14th European Conference on Applied Superconductivity (EUCAS2019) 1-5 September 2019, Glasgow, UK, pp. 012097: 1-9, 2020.

[137] K. Develos-Bagarinao, H. Yamasaki, K. Ohki and Y. Nakagawa, "Multilayered structures of (RE = rare earth)Ba2Cu3Ox films: an approach for the growth of superior quality large-area superconducting films on sapphire substrates," *Superconductor Science and Technology,* vol. 20, no. 6, pp. L25-L29, 2007.

[138] J. Metson, "2 - Production of alumina," in *Fundamentals of Aluminium Metallurgy,*

Woodhead Publishng, 2011, pp. 23-48.

[139] M. Vilardell, J. Fornell, J. Sort, R. Vlad, J. Fernández, J. Puig, A. Usoskin, A. Palau, T. Puig, X. Obradors and A. Calleja, "Inkjet-Printed Chemical Solution Y2O3 Layers for Planarization of Technical Substrates," *Coatings,* vol. 7, no. 12, pp. 227: 1-16, 2017.

[140] J. Xiong, B. Tao, W. Qin, X. Feng, X. Song, F. Zhang and Y. Li, "Reel-to-Reel Continuous Deposition of CexZr1-xO2 Single Buffer Layer for YBCO Coated Conductors," *Journal of Physics: Conference Series,* vol. 153, pp. 012036: 1-9, 2009.

[141] L. Lei, L. Li, S. Wang, G. Zhao, J. Jia, Y. Oshima, L. Zhao, L. Jin, Y. Wang, C. Li and P. Zhang, "A compositional gradient Ce1-xZrxO2 buffer architecture for producing high-performance YBCO film," *Materials Characterization,* vol. 142, pp. 383-388, 2018.

[142] L. Li, G. Zhao, L. Lei, F. Yan, J. Jia, X. Liu and Y. He, "Epitaxial growth of gradiant Ce1-xZrxO2 buffer layer for YBa2Cu3O7-x coated conductors," *Physica C: Superconductivity and its Applications,* vol. 560, pp. 26-40, 2019.

[143] M. J. Scholand and H. E. Dillon, "Life-Cycle Assessment of Energy and Environmental Impacts of LED Lighting Products - Part2: LED Manufacturing and Performance," U.S. Department of Energy, Richland, USA, 2012.

[144] International Energy Agency (IEA), "IEA World Energy Statistics and Balances," 2017. [Online]. Available: https://doi.org/10.1787/enestats-data-en. [Accessed September 2021].

[145] M. Stemmle, F. Merschel, M. Noe and A. Hobl, "AmpaCity - Advanced superconducting medium voltage system for urban area power supply," *IEEE PES T&D Conference and Exposition,* pp. 1-5, 2014.

[146] W. T. B. de Sousa, D. Kottonau, J. Bock and M. Noe, "Investigation of a Concentric Three-Phase HTS Cable Connected to an SFCL Device," *IEEE Transactions on Applied Superconductivity,* vol. 28, no. 4, pp. 5400105: 1 - 5, 2018.

[147] J. Schwartz, T. Effio, X. Liu, Q. V. Le, A. L. Mbaruku, H. J. Schneider-Muntau, T. Shen, H. Song, U. P. Trociewitz, X. Wang and H. W. Weijers, "High Field Superconducting Solenoids Via High Temperature Superconductors," *IEEE Transactions on Applied Superconductivity,* vol. 18, no. 2, pp. 70-81, 2008.

[148] P. Bruzzone, W. H. Fietz, J. V. Minervini, M. Novikov, N. Yanagi, Y. Zhai and J. Zheng, "High temperature superconductors for fusion magnets," *Nuclear Fusion,* vol. 58, no. 10, pp. 103001: 1 - 13, 2018.

[149] M. Filipenko, L. Kühn, T. Gleixner, M. Thummel, M. Lessman, D. Möller, M. Böhm, A. Schröter, K. Häse, J. Grundmann, M. Wilke, M. Frank, P. van Hasselt, J. Richter, M. Herranz-Garcia, C. Weidermann, A. Spangolo, M. Klöpzig, P.

Gröppel and S. Moldenhauer, “Concept design of a high power superconducting generator for future hybrid-electric aircraft,” *Superconductor Science and Technology,* vol. 33, no. 5, pp. 054002: 1 - 16, 2020.

[150] T. Winkler and EcoSwing Consortium, “The EcoSwing Project,” *IOP Conference Series: Materials Science and Engineering,* vol. 502, pp. 012004: 1-9, 2019.

[151] S. J. Lee, M. Park, I. Yu, Y. Won, Y. Kwak and C. Lee, “Recent Status and Progress on HTS Cables for AC and DC Power Transmission in Korea,” *IEEE Transactions on Applied Superconductivity,* vol. 28, no. 4, pp. 5401205: 1 - 5, 2018.

[152] F. Herzog, T. Kutz, M. Stemmle and T. Kugel, “Cooling unit for the AmpaCity project - One year successful operation,” *Cryogenics,* vol. 80, no. 2, pp. 204-209, 2016.

[153] T. Nakano, O. Maruyama, S. Honjo, M. Watanabe, T. Masuda, M. Hirose, M. Shimoda, N. Nakamura, H. Yaguchi and A. Machida, “Long-term operating characteristics of Japan's first in-grid HTS power cable,” *Physica C,* vol. 518, pp. 126-129, 2015.

[154] W. Yuan, S. Venuturumilli, Z. Zhang, Y. Mavrocostanti and M. Zhang, “Economic Feasibility Study of Using High-Temperature Superconducting Cables in U.K.'s Electrical Distribution Networks,” *IEEE Transactions on Applied Superconductivity,* vol. 28, no. 4, pp. 5401505: 1-5, 2018.

[155] D. Kottonau, E. Shabagin, M. Noe and S. Grohmann, “Opportunities for High-Voltage AC Superconducting Cables as Part of New Long-Distance Transmission Lines,” *IEEE Transactions on Applied Superconductivity,* vol. 27, no. 4, pp. 5400405: 1-5, 2017.

[156] C. Lee, J. Choi, H. Yang, M. Park and M. Iwakuma, “Economic Evaluation of 23 kV Tri-Axial HTS Cable Application to Power System,” *IEEE Transactions on Applied Superconductivity,* vol. 29, no. 5, pp. 5402507: 1-7, 2019.

[157] T. Yang, H. Jin, W. Li, W. Hong, B. Tian, Q. Li and Y. Xin, “Feasibility study of high temperature superconducting cables for distribution power grids in metropolises,” *Superconductor Science and Technology,* vol. 33, no. 1, pp. 014002: 1-10, 2020.

[158] M. Stemmle, F. Merschel, M. Noe, L. Hofmann and A. Hobl, “Novel Grid Concepts for Urban Area Power Supply,” *Physics Procedia,* vol. 36, pp. 884-889, 2012.

[159] M. Noe, “Superconducting Cables,” *EUCAS Short Course Power Applications, September 1st 2019, Glasgow,* 2019.

[160] M. Stemmle, F. Merschel, M. Noe and A. Hobl, “AmpaCity - Installation of Advanced Superconducting 10 kV System in City Center Replaces Conventional 110 kV Cables,” *Proceeding of 2013 IEEE International Conference on Applied*

Superconductivity and Electromagnetic Devices, pp. 323-326, 2013.

[161] H. Yumura, T. Masuda, M. Watanabe, H. Takigawa, Y. Ashibe, H. Ito, M. Hirose and K. Sato, “Albany HTS cable project long term in-grid operation status update,” *AIP Conference Proceedings,* vol. 985, pp. 1051-1058, 2008.

[162] P. Anders, J. Bock, W. Goldacker, F. Grilli, A. Hobl, L. Hofmann, M. Kleimaier, F. Merschel, M. Noe, A. Petermann, C. Prause, B. R. Oswald, K. Schippl, F. Schmidt, B. Slomka, R. Soika, M. Stemmle and T. Theisen, “Supraleitende Mittelspannungskabel zur Stromversorgung von Städten - Studie zur technischen und wirtschaftlichen Machbarkeit des Einsatzes supraleitender Kabel am Beispiel der Innenstadt von Essen (Interner Bericht),” 2011.

[163] A. Buchholz, M. Noe, D. Kottonau, E. Shabagin and M. Weil, “Environmental Life-Cycle Assessment of a 10 kV High-Temperature Superconducting Cable System for Energy Distribution,” *IEEE Transactions on Applied Superconductivity,* vol. 31, no. 5, pp. 4802405: 1-5, 2021.

[164] M. Wolf and H. Junge, “Belastungskurven und Dauerlinien in der elektrischen Energiewirtschaft,” in *Enzyklopädie der Energiewirtschaft*, vol. Zweiter Band, M. Wolf, Ed., Springer Verlag, 1959.

[165] D. Kottonau, E. Shabagin, W. T. B. de Sousa, J. Geisbüsch, M. Noe, H. Stagge, S. Fechner, H. Woiton and T. Küsters, Evaluation of the Use of Superconducting 380 kV Cable, vol. 030, M. Noe and M. Siegel, Eds., Karlsruhe: KIT Scientific Publishing - Karlsruher Schriftenreihe zur Supraleitung, 2020.

[166] DIN Deutsches Institut für Normung e.V., Ed., DIN EN 10220: 2003-03 Nahtlose und geschweißte Stahlrohre - Allgemeine Tabellen für Maße und längenbezogene Masse, Berlin: Beuth Verlag.

[167] T. Jacob, Life Cycle Assessment of Cooling Systems for a 10 kV High-Temperature Superconducting Cable System, Master's thesis, Karlsruhe Institute of Technology - Institute for Technology Assessment and Systems Analyses & TU Berlin - Institute for Energy Engineering, 2021.

[168] nkt cables, “Hoch- und Höchstspannungskabelanlagen - Kabel und Garnituren bis 550 kV,” Köln, Germany.

[169] ABB Trasmissione & Distribuzione SpA, “Environmental Product Declaration - Power Transformers 40/50 MVA (ONAN/ONAF),” Legnano, Italy, 2003.

[170] Klaus Faber AG, “Mittelspannungskabel NA2XS2Y,” Saarbrücken, Germany, 2021.

[171] A. Berger, Entwicklung supraleitender, strombegrenzender Transformatoren, vol. 003, M. Noe and M. Siegel, Eds., KIT Scientific Publishing - Karlsruher Schriftenreihe zur Supraleitung, 2011.

[172] R. C. Lee, A. Dada and S. M. Ringo, “Cryogenic Refrigeration System for HTS Cables,” *IEEE Transactions on Applied Superconductivity,* vol. 15, no. 2, pp.

1798-1801.

List of Figures

List of Tables

List of Abbreviations

CZO	Ceria-Zirconia ($Ce_{0.9}Zr_{0.1}O_2$)
GdBCO	Gadolinium Barium Copper Oxide ($GdBa_2Cu_3O_{7-\gamma}$)
HTS	High-Temperature Superconductors / Superconducting
LCA	Life Cycle Assessment
RABiTS™	Rolling-Assisted Biaxially-Textured Substrate
SFCL	Superconducting Fault Current Limiter
YBCO	Yttrium Barium Copper Oxide ($YBa_2Cu_3O_{7-\gamma}$)

List of Own Publications

Peer Reviewed Journals

A. Buchholz, M. Noe, D. Kottonau, E. Shabagin and M. Weil, "Environmental Life-Cycle Assessment of a 10 kV High-Temperature Superconducting Cable System for Energy Distribution," *IEEE Transactions on Applied Superconductivity,* vol. 31, no. 5, pp. 4802405: 1-5, 2021.

Presentations on Conferences

A. Buchholz, M. Noe, V. Große and M. Weil, "Prospective Life Cycle Assessment of REBCO High-Temperature Superconductor Tape Production," *IEEE 19th International Power Electronics and Motion Control Conference (PEMC 2021),* Online, April 25-29 2021

A. Buchholz, M. Noe, V. Große and M. Weil, "Environmental Life-Cycle Assessment of High-Temperature Superconductor Tape Production," *Energy Transition and Sustainability Conference (APEEN 2021),* Online, January 20-21 2021

A. Buchholz, M. Noe, D. Kottonau, E. Shabagin and M. Weil, "Environmental Life-Cycle Assessment of a 10 kV High-Temperature Superconductor Cable System for Energy Distribution," *Applied Superconductivity Conference (ASC 2020),* Online, October 24 – November 7 2020

A. Buchholz, M. Noe, V. Große and M. Weil, "Environmental Life-Cycle Assessment of High-Temperature Superconductor Tape Production," *14th European Conference on Applied Superconductivity (EUCAS 2019),* Glasgow, United Kingdom, September 1-5 2019

List of Supervised Student Theses

T. Jacob, Life Cycle Assessment of Cooling Systems for a 10 kV High-Temperature Superconducting Cable System, Master's thesis, Karlsruhe Institute of Technology - Institute for Technology Assessment and Systems Analyses & TU Berlin - Institute for Energy Engineering, 2021

Appendix

Appendix A: Elementary Flow List of a Copper Conductor Wire

List of all elementary flows that are used as a resource or emitted into the environment by producing one metre of copper conductor wire that is made of 17.9g of copper.

Flow	Category	Amount	Unit
Aluminium, 24% in bauxite, 11% in crude ore, in ground	Resource in ground	2.3E-06	kg
Aluminium, in ground	Resource in ground	1.5E-04	kg
Anhydrite, in ground	Resource in ground	1.8E-09	kg
Argon-40	Resource in air	1.0E-04	kg
Barite, 15% in crude ore, in ground	Resource in ground	2.0E-05	kg
Basalt, in ground	Resource in ground	1.7E-05	kg
Borax, in ground	Resource in ground	4.4E-08	kg
Bromine, 0.23% in water	Resource in water	2.8E-09	kg
Cadmium, 0.30% in sulfide, Cd 0.18%, Pb, Zn, Ag, In, in ground	Resource in ground	4.5E-06	kg
Calcite, in ground	Resource in ground	7.6E-03	kg
Carbon dioxide, in air	Resource in air	6.0E-03	kg
Carbon, organic, in soil or biomass stock	Resource in ground	1.1E-06	kg
Carnallite	Resource in water	1.2E-07	kg
Cerium, 24% in bastnasite, 2.4% in crude ore, in ground	Resource in ground	1.0E-13	kg
Chromium, 25.5% in chromite, 11.6% in crude ore, in ground	Resource in ground	2.5E-04	kg
Chrysotile, in ground	Resource in ground	8.8E-09	kg
Cinnabar, in ground	Resource in ground	3.3E-10	kg
Clay, bentonite, in ground	Resource in ground	1.6E-05	kg
Clay, unspecified, in ground	Resource in ground	2.7E-03	kg
Coal, brown, in ground	Resource in ground	6.4E-03	kg
Coal, hard, unspecified, in ground	Resource in ground	2.1E-02	kg
Cobalt, Co 5.0E-2%, in mixed ore, in ground	Resource in ground	5.1E-08	kg
Cobalt, in ground	Resource in ground	1.9E-10	kg
Colemanite, in ground	Resource in ground	9.7E-08	kg
Copper, 0.52% in sulfide, Cu 0.27% and Mo 8.2E-3% in crude ore, in ground	Resource in ground	2.2E-03	kg
Copper, 0.59% in sulfide, Cu 0.22% and Mo 8.2E-3% in crude ore, in ground	Resource in ground	1.2E-03	kg

Copper, 0.97% in sulfide, Cu 0.36% and Mo 4.1E-2% in crude ore, in ground	Resource in ground	5.9E-07	kg
Copper, 0.99% in sulfide, Cu 0.36% and Mo 8.2E-3% in crude ore, in ground	Resource in ground	3.3E-03	kg
Copper, 1.13% in sulfide, Cu 0.76% and Ni 0.76% in crude ore, in ground	Resource in ground	1.2E-04	kg
Copper, 1.18% in sulfide, Cu 0.39% and Mo 8.2E-3% in crude ore, in ground	Resource in ground	1.8E-03	kg
Copper, 1.42% in sulfide, Cu 0.81% and Mo 8.2E-3% in crude ore, in ground	Resource in ground	2.9E-04	kg
Copper, 2.19% in sulfide, Cu 1.83% and Mo 8.2E-3% in crude ore, in ground	Resource in ground	9.5E-04	kg
Copper, Cu 0.2%, in mixed ore, in ground	Resource in ground	4.3E-08	kg
Copper, Cu 0.38%, in mixed ore, in ground	Resource in ground	3.0E-03	kg
Copper, Cu 6.8E-1%, in mixed ore, in ground	Resource in ground	7.0E-07	kg
Cu, Cu 3.2E+0%, Pt 2.5E-4%, Pd 7.3E-4%, Rh 2.0E-5%, Ni 2.3E+0% in ore, in ground	Resource in ground	1.0E-04	kg
Cu, Cu 5.2E-2%, Pt 4.8E-4%, Pd 2.0E-4%, Rh 2.4E-5%, Ni 3.7E-2% in ore, in ground	Resource in ground	8.1E-07	kg
Diatomite, in ground	Resource in ground	2.2E-12	kg
Dolomite, in ground	Resource in ground	6.5E-06	kg
Energy, geothermal, converted	Resource in ground	1.5E-03	MJ
Energy, gross calorific value, in biomass	Biotic resource	6.7E-02	MJ
Energy, gross calorific value, in biomass, primary forest	Biotic resource	1.8E-05	MJ
Energy, kinetic (in wind), converted	Resource in air	8.4E-03	MJ
Energy, potential (in hydropower reservoir), converted	Resource in water	1.4E-01	MJ
Energy, solar, converted	Resource in air	1.1E-05	MJ
Europium, 0.06% in bastnasite, 0.006% in crude ore, in ground	Resource in ground	2.6E-16	kg
Feldspar, in ground	Resource in ground	4.4E-09	kg
Fish, pelagic, in ocean	Resource in water	3.3E-19	kg
Fluorine, 4.5% in apatite, 1% in crude ore, in ground	Resource in ground	6.4E-06	kg
Fluorine, 4.5% in apatite, 3% in crude ore, in ground	Resource in ground	5.2E-07	kg
Fluorspar, 92%, in ground	Resource in ground	1.0E-04	kg
Gadolinium, 0.15% in bastnasite, 0.015% in crude ore, in ground	Resource in ground	6.4E-16	kg
Gallium, 0.014% in bauxite, in ground	Resource in ground	2.8E-15	kg
Gallium, in ground	Resource in ground	4.5E-08	kg
Gangue, bauxite, in ground	Resource in ground	1.5E-03	kg
Gas, mine, off-gas, process, coal mining	Resource in ground	1.7E-04	m3
Gas, natural, in ground	Resource in ground	7.9E-03	m3
Gold, Au 1.0E-7%, in mixed ore, in ground	Resource in ground	1.1E-11	kg
Gold, Au 1.1E-4%, Ag 4.2E-3%, in ore, in ground	Resource in ground	2.9E-11	kg
Gold, Au 1.3E-4%, Ag 4.6E-5%, in ore, in ground	Resource in ground	3.8E-11	kg
Gold, Au 1.4E-4%, in ore, in ground	Resource in ground	1.1E-10	kg

Gold, Au 1.8E-4%, in mixed ore, in ground	Resource in ground	5.1E-11	kg
Gold, Au 2.1E-4%, Ag 2.1E-4%, in ore, in ground	Resource in ground	8.2E-12	kg
Gold, Au 4.3E-4%, in ore, in ground	Resource in ground	2.2E-11	kg
Gold, Au 4.9E-5%, in ore, in ground	Resource in ground	1.1E-10	kg
Gold, Au 5.4E-4%, Ag 1.5E-5%, in ore, in ground	Resource in ground	6.2E-13	kg
Gold, Au 6.7E-4%, in ore, in ground	Resource in ground	1.2E-10	kg
Gold, Au 6.8E-4%, Ag 1.5E-4%, in ore, in ground	Resource in ground	8.4E-13	kg
Gold, Au 7.1E-4%, in ore, in ground	Resource in ground	5.4E-11	kg
Gold, Au 9.7E-4%, in mixed ore, in ground	Resource in ground	7.3E-08	kg
Gold, Au 9.7E-5%, Ag 7.6E-5%, in ore, in ground	Resource in ground	3.0E-12	kg
Granite, in ground	Resource in ground	5.7E-13	kg
Gravel, in ground	Resource in ground	3.3E-02	kg
Gypsum, in ground	Resource in ground	5.6E-05	kg
Indium, 0.005% in sulfide, In 0.003%, Pb, Zn, Ag, Cd, in ground	Resource in ground	7.6E-08	kg
Iodine, 0.03% in water	Resource in water	6.0E-10	kg
Iron, 46% in ore, 25% in crude ore, in ground	Resource in ground	1.7E-03	kg
Iron, 72% in magnetite, 14% in crude ore, in ground	Resource in ground	3.3E-06	kg
Kaolinite, 24% in crude ore, in ground	Resource in ground	1.4E-05	kg
Kieserite, 25% in crude ore, in ground	Resource in ground	1.5E-08	kg
Krypton, in air	Resource in air	6.8E-18	kg
Lanthanum, 7.2% in bastnasite, 0.72% in crude ore, in ground	Resource in ground	3.1E-14	kg
Lead, 5.0% in sulfide, Pb 3.0%, Zn, Ag, Cd, In, in ground	Resource in ground	7.6E-05	kg
Lead, Pb 0.014%, in mixed ore, in ground	Resource in ground	3.6E-04	kg
Lead, Pb 3.6E-1%, in mixed ore, in ground	Resource in ground	7.7E-08	kg
Lithium, 0.15% in brine, in ground	Resource in ground	7.6E-11	kg
Magnesite, 60% in crude ore, in ground	Resource in ground	1.4E-05	kg
Manganese, 35.7% in sedimentary deposit, 14.2% in crude ore, in ground	Resource in ground	5.4E-06	kg
Metamorphous rock, graphite containing, in ground	Resource in ground	3.2E-07	kg
Molybdenum, 0.010% in sulfide, Mo 8.2E-3% and Cu 1.83% in crude ore, in ground	Resource in ground	2.7E-05	kg
Molybdenum, 0.014% in sulfide, Mo 8.2E-3% and Cu 0.81% in crude ore, in ground	Resource in ground	6.0E-06	kg
Molybdenum, 0.016% in sulfide, Mo 8.2E-3% and Cu 0.27% in crude ore, in ground	Resource in ground	5.4E-05	kg
Molybdenum, 0.022% in sulfide, Mo 8.2E-3% and Cu 0.22% in crude ore, in ground	Resource in ground	2.8E-05	kg
Molybdenum, 0.022% in sulfide, Mo 8.2E-3% and Cu 0.36% in crude ore, in ground	Resource in ground	4.5E-05	kg
Molybdenum, 0.025% in sulfide, Mo 8.2E-3% and Cu 0.39% in crude ore, in ground	Resource in ground	3.5E-05	kg
Molybdenum, 0.11% in sulfide, Mo 4.1E-2% and Cu 0.36% in crude ore, in ground	Resource in ground	6.7E-08	kg

Neodymium, 4% in bastnasite, 0.4% in crude ore, in ground	Resource in ground	1.7E-14	kg
Ni, Ni 2.3E+0%, Pt 2.5E-4%, Pd 7.3E-4%, Rh 2.0E-5%, Cu 3.2E+0% in ore, in ground	Resource in ground	7.5E-05	kg
Ni, Ni 3.7E-2%, Pt 4.8E-4%, Pd 2.0E-4%, Rh 2.4E-5%, Cu 5.2E-2% in ore, in ground	Resource in ground	1.2E-06	kg
Nickel, 1.13% in sulfide, Ni 0.76% and Cu 0.76% in crude ore, in ground	Resource in ground	2.4E-04	kg
Nickel, 1.98% in silicates, 1.04% in crude ore, in ground	Resource in ground	1.4E-04	kg
Nickel, Ni 2.5E+0%, in mixed ore, in ground	Resource in ground	2.5E-06	kg
Nitrogen	Resource in air	5.4E-03	kg
Occupation, annual crop	Land resource	3.1E-06	m2*a
Occupation, annual crop, greenhouse	Land resource	2.9E-21	m2*a
Occupation, annual crop, irrigated	Land resource	2.0E-07	m2*a
Occupation, annual crop, irrigated, intensive	Land resource	1.0E-08	m2*a
Occupation, annual crop, non-irrigated	Land resource	8.1E-08	m2*a
Occupation, annual crop, non-irrigated, extensive	Land resource	1.1E-07	m2*a
Occupation, annual crop, non-irrigated, intensive	Land resource	2.3E-05	m2*a
Occupation, construction site	Land resource	3.9E-05	m2*a
Occupation, dump site	Land resource	3.9E-03	m2*a
Occupation, forest, extensive	Land resource	3.4E-05	m2*a
Occupation, forest, intensive	Land resource	9.9E-03	m2*a
Occupation, grassland, natural (non-use)	Land resource	1.2E-05	m2*a
Occupation, industrial area	Land resource	6.0E-04	m2*a
Occupation, inland waterbody, unspecified	Land resource	5.9E-07	m2*a
Occupation, lake, artificial	Land resource	2.3E-04	m2*a
Occupation, mineral extraction site	Land resource	1.0E-03	m2*a
Occupation, pasture, man made	Land resource	-4.2E-30	m2*a
Occupation, pasture, man made, extensive	Land resource	1.2E-12	m2*a
Occupation, pasture, man made, intensive	Land resource	8.4E-09	m2*a
Occupation, permanent crop	Land resource	1.7E-06	m2*a
Occupation, permanent crop, irrigated	Land resource	3.6E-07	m2*a
Occupation, permanent crop, irrigated, intensive	Land resource	4.8E-21	m2*a
Occupation, river, artificial	Land resource	1.8E-04	m2*a
Occupation, seabed, drilling and mining	Land resource	9.1E-07	m2*a
Occupation, seabed, infrastructure	Land resource	1.1E-08	m2*a
Occupation, shrub land, sclerophyllous	Land resource	1.0E-05	m2*a
Occupation, traffic area, rail network	Land resource	2.6E-05	m2*a
Occupation, traffic area, rail/road embankment	Land resource	1.6E-04	m2*a

Occupation, traffic area, road network	Land resource	2.0E-04	m2*a
Occupation, unspecified	Land resource	2.2E-07	m2*a
Occupation, urban, discontinuously built	Land resource	5.8E-08	m2*a
Occupation, urban/industrial fallow (non-use)	Land resource	6.3E-09	m2*a
Oil, crude, in ground	Resource in ground	4.8E-03	kg
Olivine, in ground	Resource in ground	7.0E-10	kg
Oxygen	Resource in air	2.2E-03	kg
Palladium, Pd 1.6E-6%, in mixed ore, in ground	Resource in ground	1.7E-10	kg
Pd, Pd 2.0E-4%, Pt 4.8E-4%, Rh 2.4E-5%, Ni 3.7E-2%, Cu 5.2E-2% in ore, in ground	Resource in ground	2.2E-09	kg
Pd, Pd 7.3E-4%, Pt 2.5E-4%, Rh 2.0E-5%, Ni 2.3E+0%, Cu 3.2E+0% in ore, in ground	Resource in ground	2.4E-08	kg
Peat, in ground	Biotic resource	3.3E-05	kg
Perlite, in ground	Resource in ground	1.7E-08	kg
Phosphorus, 18% in apatite, 12% in crude ore, in ground	Resource in ground	2.1E-06	kg
Phosphorus, 18% in apatite, 4% in crude ore, in ground	Resource in ground	2.5E-05	kg
Platinum, Pt 4.7E-7%, in mixed ore, in ground	Resource in ground	4.9E-11	kg
Praseodymium, 0.42% in bastnasite, 0.042% in crude ore, in ground	Resource in ground	1.8E-15	kg
Pt, Pt 2.5E-4%, Pd 7.3E-4%, Rh 2.0E-5%, Ni 2.3E+0%, Cu 3.2E+0% in ore, in ground	Resource in ground	8.2E-09	kg
Pt, Pt 4.8E-4%, Pd 2.0E-4%, Rh 2.4E-5%, Ni 3.7E-2%, Cu 5.2E-2% in ore, in ground	Resource in ground	5.3E-09	kg
Rh, Rh 2.0E-5%, Pt 2.5E-4%, Pd 7.3E-4%, Ni 2.3E+0%, Cu 3.2E+0% in ore, in ground	Resource in ground	6.5E-10	kg
Rh, Rh 2.4E-5%, Pt 4.8E-4%, Pd 2.0E-4%, Ni 3.7E-2%, Cu 5.2E-2% in ore, in ground	Resource in ground	2.7E-10	kg
Rhenium, in crude ore, in ground	Resource in ground	4.0E-13	kg
Rhodium, Rh 1.6E-7%, in mixed ore, in ground	Resource in ground	1.7E-11	kg
Samarium, 0.3% in bastnasite, 0.03% in crude ore, in ground	Resource in ground	1.3E-15	kg
Sand, unspecified, in ground	Resource in ground	2.7E-07	kg
Shale, in ground	Resource in ground	6.2E-03	kg
Silver, 0.007% in sulfide, Ag 0.004%, Pb, Zn, Cd, In, in ground	Resource in ground	1.1E-07	kg
Silver, 3.2ppm in sulfide, Ag 1.2ppm, Cu and Te, in crude ore, in ground	Resource in ground	3.4E-13	kg
Silver, Ag 1.5E-4%, Au 6.8E-4%, in ore, in ground	Resource in ground	1.9E-13	kg
Silver, Ag 1.5E-5%, Au 5.4E-4%, in ore, in ground	Resource in ground	1.7E-14	kg
Silver, Ag 1.8E-6%, in mixed ore, in ground	Resource in ground	1.9E-10	kg
Silver, Ag 2.1E-4%, Au 2.1E-4%, in ore, in ground	Resource in ground	8.3E-12	kg
Silver, Ag 4.2E-3%, Au 1.1E-4%, in ore, in ground	Resource in ground	1.1E-09	kg
Silver, Ag 4.6E-5%, Au 1.3E-4%, in ore, in ground	Resource in ground	1.3E-11	kg
Silver, Ag 5.4E-3%, in mixed ore, in ground	Resource in ground	1.2E-09	kg
Silver, Ag 7.6E-5%, Au 9.7E-5%, in ore, in ground	Resource in ground	2.4E-12	kg

Silver, Ag 9.7E-4%, in mixed ore, in ground	Resource in ground	3.7E-06	kg
Sodium chloride, in ground	Resource in ground	1.8E-03	kg
Sodium nitrate, in ground	Resource in ground	3.6E-13	kg
Sodium sulphate, various forms, in ground	Resource in ground	5.0E-06	kg
Spodumene, in ground	Resource in ground	8.2E-10	kg
Stibnite, in ground	Resource in ground	2.2E-13	kg
strontium, in ground	Resource in ground	1.0E-07	kg
Sulfur, in ground	Resource in ground	4.4E-07	kg
Sylvite, 25 % in sylvinite, in ground	Resource in ground	4.4E-06	kg
Talc, in ground	Resource in ground	1.5E-06	kg
Tantalum, 81.9% in tantalite, 1.6E-4% in crude ore, in ground	Resource in ground	2.4E-09	kg
Tellurium, 0.5ppm in sulfide, Te 0.2ppm, Cu and Ag, in crude ore, in ground	Resource in ground	5.1E-14	kg
Tin, 79% in cassiterite, 0.1% in crude ore, in ground	Resource in ground	7.2E-08	kg
TiO2, 54% in ilmenite, 18% in crude ore, in ground	Resource in ground	2.5E-06	kg
TiO2, 54% in ilmenite, 2.6% in crude ore, in ground	Resource in ground	3.4E-05	kg
TiO2, 95% in rutile, 0.40% in crude ore, in ground	Resource in ground	5.3E-06	kg
Transformation, from annual crop	Land resource	4.1E-06	m2
Transformation, from annual crop, greenhouse	Land resource	5.8E-21	m2
Transformation, from annual crop, irrigated, intensive	Land resource	1.1E-25	m2
Transformation, from annual crop, non-irrigated	Land resource	3.2E-07	m2
Transformation, from annual crop, non-irrigated, extensive	Land resource	9.9E-08	m2
Transformation, from annual crop, non-irrigated, intensive	Land resource	4.0E-05	m2
Transformation, from cropland fallow (non-use)	Land resource	2.9E-08	m2
Transformation, from dump site, inert material landfill	Land resource	2.5E-07	m2
Transformation, from dump site, residual material landfill	Land resource	1.8E-06	m2
Transformation, from dump site, sanitary landfill	Land resource	2.6E-08	m2
Transformation, from dump site, slag compartment	Land resource	3.0E-08	m2
Transformation, from forest, extensive	Land resource	8.3E-06	m2
Transformation, from forest, intensive	Land resource	1.1E-04	m2
Transformation, from forest, primary (non-use)	Land resource	1.9E-07	m2
Transformation, from forest, secondary (non-use)	Land resource	4.8E-08	m2
Transformation, from forest, unspecified	Land resource	6.4E-06	m2
Transformation, from grassland, natural (non-use)	Land resource	5.2E-09	m2
Transformation, from grassland, natural, for livestock grazing	Land resource	1.4E-07	m2
Transformation, from heterogeneous, agricultural	Land resource	1.4E-09	m2

Transformation, from industrial area	Land resource	1.2E-07	m2
Transformation, from mineral extraction site	Land resource	7.7E-05	m2
Transformation, from pasture, man made	Land resource	4.5E-06	m2
Transformation, from pasture, man made, extensive	Land resource	2.4E-14	m2
Transformation, from pasture, man made, intensive	Land resource	4.3E-08	m2
Transformation, from permanent crop	Land resource	9.6E-08	m2
Transformation, from permanent crop, irrigated	Land resource	9.0E-09	m2
Transformation, from permanent crop, irrigated, intensive	Land resource	6.0E-23	m2
Transformation, from seabed, infrastructure	Land resource	8.0E-11	m2
Transformation, from seabed, unspecified	Land resource	9.2E-07	m2
Transformation, from shrub land, sclerophyllous	Land resource	2.9E-06	m2
Transformation, from traffic area, rail/road embankment	Land resource	8.1E-07	m2
Transformation, from traffic area, road network	Land resource	-2.0E-21	m2
Transformation, from unspecified	Land resource	1.2E-04	m2
Transformation, from unspecified, natural (non-use)	Land resource	3.3E-09	m2
Transformation, from wetland, inland (non-use)	Land resource	8.1E-12	m2
Transformation, to annual crop	Land resource	1.8E-06	m2
Transformation, to annual crop, greenhouse	Land resource	5.8E-21	m2
Transformation, to annual crop, irrigated, intensive	Land resource	1.0E-08	m2
Transformation, to annual crop, non-irrigated	Land resource	2.3E-07	m2
Transformation, to annual crop, non-irrigated, extensive	Land resource	1.4E-07	m2
Transformation, to annual crop, non-irrigated, intensive	Land resource	4.2E-05	m2
Transformation, to arable land, unspecified use	Land resource	1.4E-06	m2
Transformation, to cropland fallow (non-use)	Land resource	8.4E-08	m2
Transformation, to dump site	Land resource	2.9E-05	m2
Transformation, to dump site, inert material landfill	Land resource	2.5E-07	m2
Transformation, to dump site, residual material landfill	Land resource	1.8E-06	m2
Transformation, to dump site, sanitary landfill	Land resource	2.6E-08	m2
Transformation, to dump site, slag compartment	Land resource	3.0E-08	m2
Transformation, to forest, extensive	Land resource	2.6E-07	m2
Transformation, to forest, intensive	Land resource	1.2E-04	m2
Transformation, to forest, secondary (non-use)	Land resource	-2.3E-21	m2
Transformation, to forest, unspecified	Land resource	2.5E-06	m2
Transformation, to grassland, natural (non-use)	Land resource	1.6E-07	m2
Transformation, to heterogeneous, agricultural	Land resource	6.3E-07	m2

Transformation, to industrial area	Land resource	1.2E-05	m2
Transformation, to inland waterbody, unspecified	Land resource	5.9E-09	m2
Transformation, to lake, artificial	Land resource	2.0E-06	m2
Transformation, to mineral extraction site	Land resource	8.5E-05	m2
Transformation, to pasture, man made	Land resource	1.0E-07	m2
Transformation, to pasture, man made, extensive	Land resource	2.4E-14	m2
Transformation, to pasture, man made, intensive	Land resource	4.2E-10	m2
Transformation, to permanent crop	Land resource	1.1E-07	m2
Transformation, to permanent crop, irrigated	Land resource	9.0E-09	m2
Transformation, to permanent crop, irrigated, intensive	Land resource	6.0E-23	m2
Transformation, to permanent crop, non-irrigated	Land resource	-2.3E-21	m2
Transformation, to river, artificial	Land resource	2.1E-06	m2
Transformation, to seabed, drilling and mining	Land resource	9.1E-07	m2
Transformation, to seabed, infrastructure	Land resource	4.1E-09	m2
Transformation, to seabed, unspecified	Land resource	8.0E-11	m2
Transformation, to shrub land, sclerophyllous	Land resource	2.1E-06	m2
Transformation, to traffic area, rail network	Land resource	6.0E-08	m2
Transformation, to traffic area, rail/road embankment	Land resource	1.3E-06	m2
Transformation, to traffic area, road network	Land resource	1.6E-06	m2
Transformation, to unspecified	Land resource	7.5E-05	m2
Transformation, to urban, discontinuously built	Land resource	1.2E-09	m2
Transformation, to urban/industrial fallow (non-use)	Land resource	8.4E-11	m2
Transformation, to wetland, inland (non-use)	Land resource	-7.4E-21	m2
Tungsten	Resource in ground	-1.2E-20	kg
Tungsten	Unspecified resource	-1.9E-20	kg
Ulexite, in ground	Resource in ground	1.4E-08	kg
Uranium, in ground	Resource in ground	2.4E-07	kg
Volume occupied, final repository for low-active radioactive waste	Resource in ground	9.6E-10	m3
Volume occupied, final repository for radioactive waste	Resource in ground	8.2E-11	m3
Volume occupied, reservoir	Resource in water	7.1E-04	m3*a
Volume occupied, underground deposit	Resource in ground	1.7E-07	m3
Water, cooling, unspecified natural origin	Resource in water	5.3E-03	m3
Water, lake	Resource in water	3.2E-06	m3
Water, river	Resource in water	1.7E-03	m3
Water, salt, ocean	Resource in water	7.6E-05	m3

Water, salt, sole	Resource in water	2.4E-06	m3
Water, turbine use, unspecified natural origin	Resource in water	1.4E+00	m3
Water, unspecified natural origin	Resource in ground	2.0E-07	m3
Water, unspecified natural origin	Resource in water	1.1E-04	m3
Water, well, in ground	Resource in water	1.2E-04	m3
Wood, hard, standing	Biotic resource	3.2E-06	m3
Wood, soft, standing	Biotic resource	3.0E-06	m3
Wood, unspecified, standing	Biotic resource	1.7E-11	m3
Xenon, in air	Resource in air	8.0E-19	kg
Zinc, 9.0% in sulfide, Zn 5.3%, Pb, Ag, Cd, In, in ground	Resource in ground	1.4E-04	kg
Zinc, Zn 0.63%, in mixed ore, in ground	Resource in ground	4.7E-04	kg
Zinc, Zn 3.1%, in mixed ore, in ground	Resource in ground	6.6E-07	kg
Zirconium, 50% in zircon, 0.39% in crude ore, in ground	Resource in ground	5.1E-06	kg
1,3-Dioxolan-2-one	Emission to water	2.1E-09	kg
1,4-Butanediol	Emissions to air	9.3E-13	kg
1-Pentanol	Emissions to air	2.4E-13	kg
1-Pentanol	Emission to water	5.7E-13	kg
1-Pentene	Emissions to air	1.7E-12	kg
1-Pentene	Emission to water	4.3E-13	kg
2,2,4-Trimethyl pentane	Emissions to air	8.5E-17	kg
2,4-D	Emissions to air	2.8E-12	kg
2,4-D	Emissions to soil	2.0E-10	kg
2,4-D amines	Emissions to air	7.4E-21	kg
2,4-D amines	Emissions to soil	1.2E-18	kg
2,4-D amines	Emission to water	3.9E-20	kg
2,4-D ester	Emissions to air	9.8E-21	kg
2,4-D ester	Emissions to soil	1.1E-18	kg
2,4-D ester	Emission to water	3.6E-20	kg
2,4-DB	Emissions to air	4.5E-21	kg
2,4-DB	Emissions to soil	7.8E-20	kg
2,4-DB	Emission to water	2.1E-21	kg
2-Aminopropanol	Emissions to air	5.4E-15	kg
2-Aminopropanol	Emission to water	1.3E-14	kg
2-chlorobenzaldehyde	Emission to water	-1.1E-22	kg
2-Methyl pentane	Emissions to air	1.2E-10	kg

2-Methyl-1-propanol	Emissions to air	4.2E-13	kg
2-Methyl-1-propanol	Emission to water	1.0E-12	kg
2-Methyl-2-butene	Emissions to air	2.5E-16	kg
2-Methyl-2-butene	Emission to water	5.9E-16	kg
2-Nitrobenzoic acid	Emissions to air	9.7E-15	kg
2-Propanol	Emissions to air	1.3E-07	kg
2-Propanol	Emission to water	2.2E-09	kg
4-Methyl-2-pentanol	Emission to water	9.5E-18	kg
4-Methyl-2-pentanone	Emissions to air	6.8E-14	kg
4-Methyl-2-pentanone	Emission to water	1.5E-11	kg
Abamectin	Emissions to soil	2.3E-27	kg
Acenaphthene	Emissions to air	2.3E-12	kg
Acenaphthene	Emission to water	1.9E-12	kg
Acenaphthylene	Emissions to air	2.2E-13	kg
Acenaphthylene	Emission to water	7.2E-14	kg
Acephate	Emissions to air	3.0E-13	kg
Acephate	Emissions to soil	2.8E-12	kg
Acetaldehyde	Emissions to air	8.5E-08	kg
Acetaldehyde	Emission to water	8.9E-09	kg
Acetamide	Emissions to air	7.3E-14	kg
Acetamide	Emissions to soil	3.9E-13	kg
Acetic acid	Emissions to air	1.8E-07	kg
Acetic acid	Emission to water	7.8E-09	kg
Acetochlor	Emissions to soil	1.9E-13	kg
Acetone	Emissions to air	3.8E-08	kg
Acetone	Emission to water	7.9E-10	kg
Acetonitrile	Emissions to air	8.0E-11	kg
Acetonitrile	Emission to water	1.4E-14	kg
Acetyl chloride	Emission to water	4.5E-13	kg
Acidity, unspecified	Emission to water	3.1E-09	kg
Acifluorfen	Emissions to air	4.1E-14	kg
Acifluorfen	Emissions to soil	1.7E-15	kg
Aclonifen	Emissions to soil	2.4E-16	kg
Acrinathrin	Emissions to soil	3.7E-24	kg
Acrolein	Emissions to air	4.1E-09	kg

Acrylate, ion	Emission to water	6.0E-11	kg
Acrylic acid	Emission to water	2.5E-11	kg
Actinides, radioactive, unspecified	Emissions to air	4.6E-06	kBq
Actinides, radioactive, unspecified	Emission to water	1.6E-07	kBq
Aerosols, radioactive, unspecified	Emissions to air	2.2E-08	kBq
Alachlor	Emissions to air	2.9E-13	kg
Alachlor	Emissions to soil	3.6E-14	kg
Aldehydes, unspecified	Emissions to air	2.2E-08	kg
Aldicarb	Emissions to soil	9.3E-12	kg
Aldrin	Emissions to soil	2.1E-11	kg
Allyl chloride	Emission to water	6.9E-12	kg
Aluminium	Emissions to air	2.7E-04	kg
Aluminium	Emissions to soil	2.5E-06	kg
Aluminium	Emission to water	5.0E-03	kg
Aluminium hydroxide	Emission to water	3.6E-12	kg
Amidosulfuron	Emissions to soil	5.0E-16	kg
Ammonia	Emissions to air	1.1E-04	kg
Ammonium carbonate	Emissions to air	3.1E-11	kg
Ammonium, ion	Emission to water	1.6E-06	kg
Aniline	Emissions to air	9.1E-13	kg
Aniline	Emission to water	1.4E-11	kg
Anthracene	Emissions to air	3.4E-20	kg
Anthracene	Emission to water	1.7E-13	kg
Anthranilic acid	Emissions to air	7.6E-15	kg
Anthraquinone	Emissions to soil	1.6E-14	kg
Antimony	Emissions to air	8.4E-07	kg
Antimony	Emissions to soil	1.2E-09	kg
Antimony	Emission to water	1.1E-05	kg
Antimony-122	Emission to water	4.0E-09	kBq
Antimony-124	Emissions to air	5.9E-12	kBq
Antimony-124	Emission to water	1.1E-05	kBq
Antimony-125	Emissions to air	1.0E-10	kBq
Antimony-125	Emission to water	2.0E-07	kBq
AOX, Adsorbable Organic Halogen as Cl	Emission to water	2.2E-08	kg
Argon-40	Emissions to air	1.1E-06	kg

Argon-41	Emissions to air	1.2E-05	kBq
Arsenic	Emissions to air	6.4E-06	kg
Arsenic	Emissions to soil	3.6E-10	kg
Arsenic, ion	Emission to water	3.4E-05	kg
Arsine	Emissions to air	2.9E-16	kg
Asulam	Emissions to soil	3.0E-18	kg
Atrazine	Emissions to air	2.3E-13	kg
Atrazine	Emissions to soil	4.3E-11	kg
Atrazine	Emission to water	1.2E-17	kg
Azoxystrobin	Emissions to air	1.3E-13	kg
Azoxystrobin	Emissions to soil	1.8E-13	kg
Barite	Emission to water	1.1E-06	kg
Barium	Emissions to air	4.5E-08	kg
Barium	Emissions to soil	8.0E-08	kg
Barium	Emission to water	1.1E-05	kg
Barium sulfide	Emission to water	-5.7E-21	kg
Barium-140	Emissions to air	3.3E-09	kBq
Barium-140	Emission to water	8.5E-09	kBq
Benomyl	Emissions to soil	1.6E-13	kg
Bensulfuron methyl ester	Emissions to soil	1.6E-14	kg
Bentazone	Emissions to air	1.3E-13	kg
Bentazone	Emissions to soil	1.4E-13	kg
Bentazone	Emission to water	2.6E-15	kg
Benz(a)anthracene	Emissions to air	4.3E-15	kg
Benz(a)anthracene	Emission to water	6.4E-16	kg
Benzal chloride	Emissions to air	4.4E-15	kg
Benzaldehyde	Emissions to air	3.1E-09	kg
Benzene	Emissions to air	1.0E-06	kg
Benzene	Emission to water	1.2E-07	kg
Benzene, chloro-	Emission to water	2.4E-09	kg
Benzene, dichloro	Emissions to air	4.6E-13	kg
Benzene, ethyl-	Emissions to air	5.1E-08	kg
Benzene, ethyl-	Emission to water	4.7E-09	kg
Benzene, hexachloro-	Emissions to air	9.2E-12	kg
Benzene, pentachloro-	Emissions to air	1.5E-13	kg

Benzo(a)pyrene	Emissions to air	3.3E-09	kg
Benzo(a)pyrene	Emission to water	7.8E-17	kg
Benzo(b)fluoranthene	Emissions to air	5.1E-15	kg
Benzo(b)fluoranthene	Emission to water	7.6E-17	kg
Benzo(ghi)perylene	Emissions to air	3.1E-16	kg
Benzo(ghi)perylene	Emission to water	1.1E-17	kg
Benzo(k)fluoranthene	Emissions to air	3.7E-15	kg
Benzo(k)fluoranthene	Emission to water	3.6E-17	kg
Benzyl alcohol	Emission to water	2.1E-22	kg
Beryllium	Emissions to air	1.5E-09	kg
Beryllium	Emissions to soil	9.4E-11	kg
Beryllium	Emission to water	1.3E-08	kg
Bifenox	Emissions to soil	1.1E-14	kg
Bifenthrin	Emissions to soil	6.9E-16	kg
Bisphenol A	Emission to water	6.8E-10	kg
Bitertanol	Emissions to soil	2.3E-16	kg
BOD5, Biological Oxygen Demand	Emission to water	1.9E-04	kg
Borate	Emission to water	2.3E-10	kg
Boric acid	Emissions to air	1.6E-19	kg
Boron	Emissions to air	2.7E-07	kg
Boron	Emissions to soil	5.1E-09	kg
Boron	Emission to water	9.4E-04	kg
Boron trifluoride	Emissions to air	1.1E-15	kg
Boscalid	Emissions to soil	1.0E-22	kg
Bromate	Emission to water	6.1E-08	kg
Bromide	Emission to water	2.4E-09	kg
Bromine	Emissions to air	8.1E-08	kg
Bromine	Emissions to soil	7.2E-10	kg
Bromine	Emission to water	1.6E-06	kg
Bromoxynil	Emissions to air	4.5E-20	kg
Bromoxynil	Emissions to soil	4.2E-14	kg
Bromoxynil	Emission to water	1.3E-20	kg
Bromuconazole	Emissions to soil	7.1E-16	kg
Butadiene	Emissions to air	1.6E-13	kg
Butane	Emissions to air	4.5E-07	kg

Butanol	Emissions to air	1.3E-12	kg
Butanol	Emission to water	1.8E-09	kg
Butene	Emissions to air	4.9E-09	kg
Butene	Emission to water	1.4E-10	kg
Butyl acetate	Emission to water	2.4E-09	kg
Butyrolactone	Emission to water	1.1E-13	kg
Cadmium	Emissions to air	2.2E-06	kg
Cadmium	Emissions to soil	8.6E-10	kg
Cadmium, ion	Emission to water	2.2E-05	kg
Calcium	Emissions to air	6.2E-07	kg
Calcium	Emissions to soil	2.9E-06	kg
Calcium, ion	Emission to water	3.7E-02	kg
Captan	Emissions to soil	4.2E-23	kg
Carbaryl	Emissions to air	3.4E-14	kg
Carbaryl	Emissions to soil	2.4E-14	kg
Carbaryl	Emission to water	1.5E-21	kg
Carbendazim	Emissions to soil	1.4E-12	kg
Carbetamide	Emissions to soil	1.6E-14	kg
Carbofuran	Emissions to soil	8.5E-11	kg
Carbon	Emissions to soil	9.4E-06	kg
Carbon dioxide	Emissions to air	1.6E-22	kg
Carbon dioxide, fossil	Emissions to air	7.3E-02	kg
Carbon dioxide, from soil or biomass stock	Emissions to air	1.0E-04	kg
Carbon dioxide, non-fossil	Emissions to air	4.4E-03	kg
Carbon dioxide, to soil or biomass stock	Emissions to soil	3.4E-08	kg
Carbon disulfide	Emissions to air	1.1E-04	kg
Carbon disulfide	Emission to water	2.8E-11	kg
Carbon monoxide, fossil	Emissions to air	2.6E-04	kg
Carbon monoxide, from soil or biomass stock	Emissions to air	4.6E-08	kg
Carbon monoxide, non-fossil	Emissions to air	1.3E-04	kg
Carbon-14	Emissions to air	3.1E-04	kBq
Carbon-14	Emission to water	1.2E-06	kBq
Carbonate	Emission to water	4.1E-08	kg
Carbonyl sulfide	Emissions to air	7.1E-08	kg
Carboxylic acids, unspecified	Emission to water	7.7E-07	kg

Carfentrazone ethyl ester	Emissions to soil	3.8E-16	kg
Carfentrazone-ethyl	Emissions to air	3.7E-15	kg
Cerium-141	Emissions to air	7.9E-10	kBq
Cerium-141	Emission to water	3.8E-09	kBq
Cerium-144	Emission to water	2.1E-09	kBq
Cesium	Emission to water	1.8E-10	kg
Cesium-134	Emissions to air	3.8E-11	kBq
Cesium-134	Emission to water	1.0E-07	kBq
Cesium-136	Emission to water	1.2E-09	kBq
Cesium-137	Emissions to air	6.9E-10	kBq
Cesium-137	Emission to water	1.8E-05	kBq
Cesium-137	Emission to water	1.3E-06	kBq
Chloramine	Emissions to air	1.4E-12	kg
Chloramine	Emission to water	1.2E-11	kg
Chlorate	Emission to water	6.3E-07	kg
Chlorfenvinphos	Emissions to soil	1.1E-30	kg
Chloridazon	Emissions to soil	6.3E-14	kg
Chloride	Emissions to soil	9.4E-07	kg
Chloride	Emission to water	6.7E-04	kg
Chloride, ion	Emission to water	2.5E-09	kg
Chlorides, unspecified	Emission to water	4.9E-06	kg
Chlorimuron-ethyl	Emissions to air	6.8E-14	kg
Chlorimuron-ethyl	Emissions to soil	7.0E-14	kg
Chlorinated solvents, unspecified	Emissions to air	2.5E-12	kg
Chlorinated solvents, unspecified	Emission to water	6.0E-09	kg
Chlorine	Emissions to air	1.2E-07	kg
Chlorine	Emissions to soil	1.0E-09	kg
Chlorine	Emission to water	4.5E-09	kg
Chlormequat	Emissions to soil	1.0E-12	kg
Chloroacetic acid	Emissions to air	2.7E-12	kg
Chloroacetic acid	Emission to water	9.0E-11	kg
Chloroacetyl chloride	Emission to water	1.7E-14	kg
Chloroform	Emissions to air	2.8E-10	kg
Chloroform	Emission to water	3.6E-12	kg
Chloropicrin	Emissions to soil	3.9E-20	kg

Chlorosilane, trimethyl-	Emissions to air	1.5E-11	kg
Chlorosulfonic acid	Emissions to air	2.2E-14	kg
Chlorosulfonic acid	Emission to water	5.2E-14	kg
Chlorothalonil	Emissions to soil	6.3E-09	kg
Chlorotoluron	Emissions to soil	1.5E-14	kg
Chlorpyrifos	Emissions to air	1.4E-12	kg
Chlorpyrifos	Emissions to soil	7.5E-12	kg
Chlorpyrifos methyl	Emissions to soil	6.0E-11	kg
Chlorsulfuron	Emissions to soil	9.8E-16	kg
Choline chloride	Emissions to soil	1.3E-13	kg
Chromium	Emissions to air	9.1E-07	kg
Chromium	Emissions to soil	1.7E-08	kg
Chromium IV	Emissions to air	1.9E-16	kg
Chromium VI	Emissions to air	2.2E-08	kg
Chromium VI	Emissions to soil	5.8E-09	kg
Chromium VI	Emission to water	7.3E-06	kg
Chromium, ion	Emission to water	1.7E-08	kg
Chromium-51	Emissions to air	5.1E-11	kBq
Chromium-51	Emission to water	6.5E-07	kBq
Chrysene	Emissions to air	4.7E-16	kg
Chrysene	Emission to water	4.1E-16	kg
Cinidon-ethyl	Emissions to soil	6.0E-16	kg
Clethodim	Emissions to air	2.0E-13	kg
Clethodim	Emissions to soil	1.7E-13	kg
Clodinafop-propargyl	Emissions to soil	1.4E-14	kg
Clomazone	Emissions to soil	7.5E-13	kg
Clopyralid	Emissions to soil	2.1E-14	kg
Cloquintocet-mexyl	Emissions to soil	3.5E-15	kg
Cloransulam-methyl	Emissions to air	3.5E-14	kg
Cloransulam-methyl	Emissions to soil	3.0E-14	kg
Cobalt	Emissions to air	2.6E-07	kg
Cobalt	Emissions to soil	5.9E-10	kg
Cobalt	Emission to water	6.0E-05	kg
Cobalt-57	Emission to water	3.9E-08	kBq
Cobalt-58	Emissions to air	1.1E-10	kBq

Cobalt-58	Emission to water	5.3E-06	kBq
Cobalt-60	Emissions to air	8.0E-10	kBq
Cobalt-60	Emission to water	3.4E-06	kBq
COD, Chemical Oxygen Demand	Emission to water	5.3E-04	kg
Copper	Emissions to air	1.8E-05	kg
Copper	Emissions to soil	4.0E-08	kg
Copper, ion	Emission to water	1.7E-04	kg
Cu-HDO	Emission to water	7.0E-16	kg
Cumene	Emissions to air	1.6E-08	kg
Cumene	Emission to water	5.8E-08	kg
Cyanide	Emissions to air	6.7E-08	kg
Cyanide	Emission to water	4.3E-06	kg
Cyanoacetic acid	Emissions to air	1.7E-14	kg
Cyclohexane	Emissions to air	5.6E-18	kg
Cyclohexane (for all cycloalkanes)	Emissions to air	7.0E-12	kg
Cycloxydim	Emissions to soil	1.7E-24	kg
Cyfluthrin	Emissions to air	7.1E-15	kg
Cyfluthrin	Emissions to soil	2.3E-14	kg
Cyhalothrin	Emissions to soil	1.8E-26	kg
Cyhalothrin, gamma-	Emissions to air	8.2E-14	kg
Cyhalothrin, gamma-	Emissions to soil	3.5E-15	kg
Cypermethrin	Emissions to soil	4.0E-11	kg
Cyproconazole	Emissions to soil	1.3E-14	kg
Cyprodinil	Emissions to soil	2.9E-13	kg
Deltamethrin	Emissions to soil	2.2E-14	kg
Desmedipham	Emissions to soil	4.1E-16	kg
Dibenz(a,h)anthracene	Emissions to air	2.4E-15	kg
Dibenz(a,h)anthracene	Emission to water	7.5E-18	kg
Dibutyltin	Emission to water	1.5E-30	kg
Dicamba	Emissions to air	2.3E-14	kg
Dicamba	Emissions to soil	1.6E-14	kg
Dicamba	Emission to water	1.3E-18	kg
Dichlorodimethylsilane	Emissions to air	1.4E-22	kg
Dichlorprop	Emissions to air	7.2E-21	kg
Dichlorprop	Emissions to soil	1.1E-18	kg

Dichlorprop	Emission to water	3.6E-20	kg
Dichlorprop-P	Emissions to soil	2.3E-14	kg
Dichromate	Emission to water	2.1E-12	kg
Dichromate	Emission to water	1.7E-10	kg
Diclofop	Emissions to soil	2.1E-14	kg
Diclofop-methyl	Emissions to soil	2.4E-14	kg
Dicrotophos	Emissions to soil	5.1E-13	kg
Diethanolamine	Emission to water	8.3E-12	kg
Diethyl ether	Emissions to air	1.3E-17	kg
Diethylamine	Emissions to air	4.1E-13	kg
Diethylamine	Emission to water	9.8E-13	kg
Diethylene glycol	Emissions to air	1.1E-17	kg
Diethylene glycol	Emission to water	1.8E-21	kg
Difenoconazole	Emissions to soil	2.1E-12	kg
Diflubenzuron	Emissions to air	3.7E-15	kg
Diflubenzuron	Emissions to soil	1.0E-10	kg
Diflufenican	Emissions to soil	4.5E-14	kg
Diflufenzopyr-sodium	Emissions to soil	6.1E-16	kg
Diisobutyl ketone	Emission to water	-4.6E-23	kg
Dimethachlor	Emissions to soil	1.2E-12	kg
Dimethenamid	Emissions to air	3.6E-19	kg
Dimethenamid	Emissions to soil	1.9E-14	kg
Dimethenamid	Emission to water	1.3E-19	kg
Dimethoate	Emissions to soil	5.3E-14	kg
Dimethyl carbonate	Emissions to air	2.0E-10	kg
Dimethyl malonate	Emissions to air	2.1E-14	kg
Dimethylamine	Emissions to air	1.7E-14	kg
Dimethylamine	Emission to water	6.4E-13	kg
Dimethyldichlorosilane	Emissions to air	1.2E-22	kg
Dimethyldichlorosilane	Emission to water	4.3E-24	kg
Dinitrogen monoxide	Emissions to air	1.6E-05	kg
Dinitrogen tetroxide	Emissions to air	3.1E-12	kg
Dioxins, measured as 2,3,7,8-tetrachlorodibenzo-p-dioxin	Emissions to air	2.1E-13	kg
Dioxins, measured as 2,3,7,8-tetrachlorodibenzo-p-dioxin	Emissions to soil	7.8E-15	kg
Diphenylether-compound	Emission to water	-1.2E-22	kg

Diphenyltin	Emission to water	9.7E-29	kg
Dipropylamine	Emissions to air	2.6E-13	kg
Dipropylamine	Emission to water	6.2E-13	kg
Diquat	Emissions to soil	2.7E-14	kg
Discarded fish, pelagic, to ocean	Emission to water	1.2E-20	kg
Dissolved solids	Emission to water	1.7E-04	kg
Dithianon	Emissions to soil	5.0E-15	kg
Diuron	Emissions to soil	5.2E-12	kg
DOC, Dissolved Organic Carbon	Emission to water	3.0E-04	kg
Elemental carbon	Emissions to air	1.1E-10	kg
Elemental carbon	Emissions to soil	3.7E-10	kg
Elemental carbon	Emission to water	3.7E-10	kg
Endosulfan	Emissions to soil	3.1E-11	kg
Endothall	Emissions to soil	1.2E-14	kg
Epichlorohydrin	Emission to water	3.0E-10	kg
Epoxiconazole	Emissions to soil	1.5E-14	kg
Esfenvalerate	Emissions to air	4.3E-14	kg
Esfenvalerate	Emissions to soil	1.9E-15	kg
Ethalfluralin	Emissions to soil	4.1E-13	kg
Ethane	Emissions to air	6.0E-06	kg
Ethane, 1,1,1,2-tetrafluoro-, HFC-134a	Emissions to air	5.2E-11	kg
Ethane, 1,1,1-trichloro-, HCFC-140	Emissions to air	4.4E-11	kg
Ethane, 1,1,1-trichloro-, HCFC-140	Emission to water	2.6E-21	kg
Ethane, 1,1,2-trichloro-1,2,2-trifluoro-, CFC-113	Emissions to air	7.9E-12	kg
Ethane, 1,1-difluoro-, HFC-152a	Emissions to air	6.7E-09	kg
Ethane, 1,2-dichloro-	Emissions to air	3.7E-07	kg
Ethane, 1,2-dichloro-	Emission to water	1.8E-09	kg
Ethane, 1,2-dichloro-1,1,2,2-tetrafluoro-, CFC-114	Emissions to air	8.4E-10	kg
Ethane, 2-chloro-1,1,1,2-tetrafluoro-, HCFC-124	Emissions to air	4.8E-12	kg
Ethane, hexafluoro-, HFC-116	Emissions to air	7.9E-10	kg
Ethanol	Emissions to air	2.4E-08	kg
Ethanol	Emission to water	4.4E-09	kg
Ethene	Emissions to air	2.2E-06	kg
Ethene	Emission to water	3.5E-08	kg
Ethene, chloro-	Emissions to air	1.5E-07	kg

Ethene, chloro-	Emission to water	2.2E-09	kg
Ethene, tetrachloro-	Emissions to air	9.6E-11	kg
Ethephon	Emissions to air	4.0E-21	kg
Ethephon	Emissions to soil	1.3E-12	kg
Ethephon	Emission to water	2.7E-22	kg
Ethofumesate	Emissions to soil	4.3E-14	kg
Ethyl acetate	Emissions to air	1.4E-08	kg
Ethyl acetate	Emission to water	3.6E-12	kg
Ethyl cellulose	Emissions to air	2.8E-11	kg
Ethylamine	Emissions to air	5.4E-13	kg
Ethylamine	Emission to water	1.3E-12	kg
Ethylene	Emissions to air	1.7E-07	kg
Ethylene diamine	Emissions to air	5.5E-12	kg
Ethylene diamine	Emission to water	1.3E-11	kg
Ethylene oxide	Emissions to air	5.9E-10	kg
Ethylene oxide	Emission to water	1.6E-10	kg
Ethyne	Emissions to air	6.2E-08	kg
Fenbuconazole	Emissions to soil	1.3E-15	kg
Fenoxaprop	Emissions to air	5.6E-14	kg
Fenoxaprop	Emissions to soil	6.1E-14	kg
Fenoxaprop ethyl ester	Emissions to soil	1.8E-15	kg
Fenoxaprop-P ethyl ester	Emissions to soil	1.4E-16	kg
Fenpiclonil	Emissions to soil	2.5E-10	kg
Fenpropidin	Emissions to soil	6.8E-14	kg
Fenpropimorph	Emissions to soil	5.4E-14	kg
Fipronil	Emissions to soil	3.0E-12	kg
Florasulam	Emissions to soil	6.6E-17	kg
Fluazifop-p-butyl	Emissions to air	8.0E-14	kg
Fluazifop-P-butyl	Emissions to soil	3.2E-13	kg
Flucarbazone sodium salt	Emissions to soil	6.1E-17	kg
Fludioxonil	Emissions to soil	1.6E-14	kg
Flufenacet	Emissions to air	3.0E-14	kg
Flufenacet	Emissions to soil	7.6E-15	kg
Flumetsulam	Emissions to air	7.0E-15	kg
Flumetsulam	Emissions to soil	1.4E-15	kg

Flumiclorac-pentyl	Emissions to air	1.2E-14	kg
Flumiclorac-pentyl	Emissions to soil	5.1E-16	kg
Flumioxazin	Emissions to air	1.2E-13	kg
Flumioxazin	Emissions to soil	3.9E-14	kg
Fluoranthene	Emissions to air	3.9E-14	kg
Fluoranthene	Emission to water	3.4E-12	kg
Fluorene	Emissions to air	3.6E-14	kg
Fluorene	Emission to water	1.2E-12	kg
Fluoride	Emissions to soil	1.4E-08	kg
Fluoride	Emission to water	1.7E-03	kg
Fluorine	Emissions to air	4.8E-07	kg
Fluosilicic acid	Emissions to air	4.6E-09	kg
Fluosilicic acid	Emission to water	8.9E-09	kg
Flupyrsulfuron-methyl	Emissions to soil	9.5E-17	kg
Fluquinconazole	Emissions to soil	1.1E-15	kg
Flurochloridone	Emissions to soil	-1.7E-26	kg
Fluroxypyr	Emissions to soil	4.3E-14	kg
Flurtamone	Emissions to soil	2.0E-14	kg
Flusilazole	Emissions to soil	4.5E-15	kg
Folpet	Emissions to soil	1.1E-26	kg
Fomesafen	Emissions to air	4.5E-13	kg
Fomesafen	Emissions to soil	2.4E-13	kg
Foramsulfuron	Emissions to soil	1.1E-16	kg
Formaldehyde	Emissions to air	2.5E-07	kg
Formaldehyde	Emission to water	4.9E-09	kg
Formamide	Emissions to air	4.4E-13	kg
Formamide	Emission to water	1.1E-12	kg
Formate	Emission to water	3.9E-11	kg
Formic acid	Emissions to air	5.1E-10	kg
Formic acid	Emission to water	3.1E-13	kg
Fresh water (obsolete)	Emission to water	1.6E-07	m3
Fungicides, unspecified	Emissions to soil	2.2E-16	kg
Furan	Emissions to air	2.1E-09	kg
Glufosinate	Emissions to soil	4.8E-12	kg
Glutaraldehyde	Emission to water	7.0E-11	kg

Glyphosate	Emissions to air	9.0E-11	kg
Glyphosate	Emissions to soil	2.0E-09	kg
Glyphosate	Emission to water	6.4E-14	kg
Halosulfuron-methyl	Emissions to soil	4.6E-15	kg
Heat, waste	Emissions to air	3.1E-03	MJ
Heat, waste	Emissions to soil	5.1E-04	MJ
Heat, waste	Emission to water	7.4E-03	MJ
Helium	Emissions to air	7.6E-09	kg
Heptane	Emissions to air	5.1E-08	kg
Herbicides, unspecified	Emissions to soil	1.9E-13	kg
Hexaconazole	Emissions to soil	6.7E-23	kg
Hexane	Emissions to air	2.4E-07	kg
Hydrazine	Emission to water	-2.2E-22	kg
Hydrocarbons, aliphatic, alkanes, cyclic	Emissions to air	1.6E-08	kg
Hydrocarbons, aliphatic, alkanes, unspecified	Emissions to air	3.4E-07	kg
Hydrocarbons, aliphatic, alkanes, unspecified	Emission to water	2.3E-08	kg
Hydrocarbons, aliphatic, unsaturated	Emissions to air	1.7E-07	kg
Hydrocarbons, aliphatic, unsaturated	Emission to water	2.2E-09	kg
Hydrocarbons, aromatic	Emissions to air	1.9E-07	kg
Hydrocarbons, aromatic	Emission to water	9.7E-08	kg
Hydrocarbons, chlorinated	Emissions to air	2.6E-09	kg
Hydrocarbons, unspecified	Emissions to air	1.2E-10	kg
Hydrocarbons, unspecified	Emissions to soil	1.2E-10	kg
Hydrocarbons, unspecified	Emission to water	1.9E-07	kg
Hydrochloric acid	Emissions to air	5.1E-13	kg
Hydrochloric acid	Emission to water	4.8E-07	kg
Hydrogen	Emissions to air	2.9E-06	kg
Hydrogen carbonate	Emission to water	2.7E-08	kg
Hydrogen chloride	Emissions to air	1.3E-05	kg
Hydrogen fluoride	Emissions to air	1.1E-06	kg
Hydrogen peroxide	Emissions to air	2.1E-11	kg
Hydrogen peroxide	Emission to water	7.0E-10	kg
Hydrogen sulfide	Emissions to air	2.3E-07	kg
Hydrogen sulfide	Emission to water	9.5E-07	kg
Hydrogen-3, Tritium	Emissions to air	8.2E-04	kBq

Hydrogen-3, Tritium	Emission to water	1.1E-01	kBq
Hydroxide	Emission to water	1.3E-09	kg
Hypochlorite	Emission to water	1.9E-08	kg
Imazamox	Emissions to air	1.8E-14	kg
Imazamox	Emissions to soil	3.0E-14	kg
Imazapyr	Emissions to soil	1.5E-17	kg
Imazaquin	Emissions to air	5.7E-14	kg
Imazaquin	Emissions to soil	2.4E-15	kg
Imazethapyr	Emissions to air	1.2E-13	kg
Imazethapyr	Emissions to soil	7.7E-14	kg
Imidacloprid	Emissions to soil	3.0E-12	kg
Indeno(1,2,3-cd)pyrene	Emissions to air	9.4E-16	kg
Indeno(1,2,3-cd)pyrene	Emission to water	1.2E-16	kg
Indoxacarb	Emissions to soil	1.5E-24	kg
Insecticides, unspecified	Emissions to soil	2.6E-20	kg
Iodide	Emissions to soil	3.4E-13	kg
Iodide	Emission to water	3.0E-08	kg
Iodine	Emissions to air	4.2E-08	kg
Iodine-129	Emissions to air	9.8E-08	kBq
Iodine-131	Emissions to air	2.8E-06	kBq
Iodine-131	Emission to water	2.2E-06	kBq
Iodine-133	Emissions to air	8.0E-09	kBq
Iodine-133	Emission to water	6.3E-09	kBq
Iodosulfuron	Emissions to soil	7.5E-17	kg
Iodosulfuron-methyl-sodium	Emissions to soil	5.9E-17	kg
Ioxynil	Emissions to soil	6.9E-14	kg
Iprodion	Emissions to soil	5.4E-13	kg
Iron	Emissions to air	7.9E-07	kg
Iron	Emissions to soil	3.0E-06	kg
Iron	Emission to water	3.3E-09	kg
Iron, ion	Emission to water	8.0E-03	kg
Iron-59	Emission to water	1.0E-05	kBq
Isocyanic acid	Emissions to air	9.8E-09	kg
Isoprene	Emissions to air	7.1E-12	kg
Isopropylamine	Emissions to air	1.2E-13	kg

Isopropylamine	Emission to water	3.0E-13	kg
Isoproturon	Emissions to soil	3.1E-13	kg
Isoxaflutole	Emissions to soil	3.5E-15	kg
Kresoxim-methyl	Emissions to soil	1.0E-14	kg
Krypton-85	Emissions to air	3.9E-05	kBq
Krypton-85m	Emissions to air	7.0E-05	kBq
Krypton-87	Emissions to air	1.1E-05	kBq
Krypton-88	Emissions to air	1.4E-05	kBq
Krypton-89	Emissions to air	5.9E-06	kBq
Lactic acid	Emissions to air	2.0E-13	kg
Lactic acid	Emission to water	4.8E-13	kg
Lactofen	Emissions to air	5.8E-14	kg
Lactofen	Emissions to soil	2.5E-15	kg
Lambda-cyhalothrin	Emissions to air	1.3E-22	kg
Lambda-cyhalothrin	Emissions to soil	8.4E-14	kg
Lambda-cyhalothrin	Emission to water	6.4E-26	kg
Lanthanum-140	Emissions to air	2.8E-10	kBq
Lanthanum-140	Emission to water	1.0E-08	kBq
Lauric acid	Emissions to air	5.4E-23	kg
Lauric acid	Emission to water	1.8E-11	kg
Lead	Emissions to air	1.7E-05	kg
Lead	Emissions to soil	3.3E-08	kg
Lead	Emission to water	1.7E-05	kg
Lead-210	Emissions to air	2.2E-05	kBq
Lead-210	Emission to water	3.3E-06	kBq
Lenacil	Emissions to soil	2.3E-16	kg
Linuron	Emissions to soil	1.8E-11	kg
Lithium	Emissions to air	1.8E-14	kg
Lithium	Emissions to soil	2.5E-12	kg
Lithium	Emission to water	2.5E-12	kg
Lithium, ion	Emission to water	3.9E-06	kg
m-Xylene	Emissions to air	4.3E-09	kg
m-Xylene	Emission to water	1.1E-10	kg
Magnesium	Emissions to air	7.3E-07	kg
Magnesium	Emissions to soil	5.4E-07	kg

Magnesium	Emission to water	2.3E-02	kg
Malathion	Emissions to soil	2.9E-12	kg
Mancozeb	Emissions to soil	8.2E-09	kg
Manganese	Emissions to air	2.4E-06	kg
Manganese	Emissions to soil	1.1E-07	kg
Manganese	Emission to water	2.5E-03	kg
Manganese-54	Emissions to air	2.6E-11	kBq
Manganese-54	Emission to water	1.9E-07	kBq
MCPA	Emissions to air	2.1E-20	kg
MCPA	Emissions to soil	3.5E-13	kg
MCPA	Emission to water	4.4E-20	kg
MCPB	Emissions to air	2.1E-20	kg
MCPB	Emissions to soil	2.3E-16	kg
MCPB	Emission to water	4.4E-20	kg
Mecoprop	Emissions to soil	5.6E-14	kg
Mecoprop-P	Emissions to soil	4.1E-14	kg
Mefenpyr	Emissions to soil	3.8E-15	kg
Mefenpyr-diethyl	Emissions to soil	6.1E-20	kg
Mepiquat chloride	Emissions to soil	7.0E-14	kg
Mercury	Emissions to air	1.8E-08	kg
Mercury	Emissions to soil	6.6E-11	kg
Mercury	Emission to water	6.1E-08	kg
Mesosulfuron-methyl (prop)	Emissions to soil	3.3E-16	kg
Mesotrione	Emissions to soil	5.0E-15	kg
Metalaxil	Emissions to soil	3.5E-13	kg
Metalaxyl-M	Emissions to soil	1.6E-22	kg
Metaldehyde	Emissions to soil	9.3E-13	kg
Metam-sodium	Emissions to soil	9.1E-13	kg
Metamitron	Emissions to soil	1.2E-14	kg
Metazachlor	Emissions to soil	2.9E-12	kg
Metconazole	Emissions to soil	1.2E-13	kg
Methane	Emissions to air	3.9E-10	kg
Methane, bromo-, Halon 1001	Emissions to air	1.0E-15	kg
Methane, bromochlorodifluoro-, Halon 1211	Emissions to air	1.8E-10	kg
Methane, bromotrifluoro-, Halon 1301	Emissions to air	2.1E-10	kg

Methane, chlorodifluoro-, HCFC-22	Emissions to air	1.8E-09	kg
Methane, dichloro-, HCC-30	Emissions to air	8.1E-10	kg
Methane, dichloro-, HCC-30	Emission to water	2.9E-09	kg
Methane, dichlorodifluoro-, CFC-12	Emissions to air	3.3E-12	kg
Methane, dichlorofluoro-, HCFC-21	Emissions to air	3.1E-14	kg
Methane, fossil	Emissions to air	1.9E-04	kg
Methane, from soil or biomass stock	Emissions to air	3.2E-09	kg
Methane, monochloro-, R-40	Emissions to air	1.2E-09	kg
Methane, non-fossil	Emissions to air	4.0E-07	kg
Methane, tetrachloro-, R-10	Emissions to air	5.0E-10	kg
Methane, tetrafluoro-, R-14	Emissions to air	1.2E-08	kg
Methane, trichlorofluoro-, CFC-11	Emissions to air	3.4E-14	kg
Methane, trifluoro-, HFC-23	Emissions to air	9.9E-12	kg
Methanesulfonic acid	Emissions to air	1.7E-14	kg
Methanol	Emissions to air	1.5E-07	kg
Methanol	Emission to water	5.7E-09	kg
Methomyl	Emissions to air	1.4E-20	kg
Methomyl	Emissions to soil	4.3E-20	kg
Methomyl	Emission to water	2.1E-22	kg
Methoxyfenozide	Emissions to soil	3.6E-26	kg
Methyl acetate	Emissions to air	2.3E-15	kg
Methyl acetate	Emission to water	5.4E-15	kg
Methyl acrylate	Emissions to air	2.9E-11	kg
Methyl acrylate	Emission to water	5.6E-10	kg
Methyl amine	Emissions to air	2.1E-14	kg
Methyl amine	Emission to water	5.0E-14	kg
Methyl borate	Emissions to air	1.5E-13	kg
Methyl ethyl ketone	Emissions to air	1.4E-08	kg
Methyl ethyl ketone	Emissions to air	8.1E-13	kg
Methyl formate	Emissions to air	1.9E-13	kg
Methyl formate	Emission to water	7.7E-14	kg
Methyl lactate	Emissions to air	2.2E-13	kg
Methyl parathion	Emissions to air	4.6E-14	kg
Methyl parathion	Emissions to soil	2.0E-15	kg
Methyl pentane	Emission to water	-8.6E-24	kg

Methylamine	Emissions to air	2.3E-13	kg
Metolachlor	Emissions to air	9.4E-13	kg
Metolachlor	Emissions to soil	1.3E-10	kg
Metolachlor	Emission to water	3.8E-16	kg
Metosulam	Emissions to soil	1.9E-16	kg
Metribuzin	Emissions to air	3.7E-13	kg
Metribuzin	Emissions to soil	2.9E-10	kg
Metsulfuron-methyl	Emissions to soil	1.2E-12	kg
Molinate	Emissions to soil	1.4E-12	kg
Molybdenum	Emissions to air	5.1E-09	kg
Molybdenum	Emissions to soil	2.7E-10	kg
Molybdenum	Emission to water	2.1E-05	kg
Molybdenum-99	Emission to water	3.1E-09	kBq
Monobutyltin	Emission to water	2.6E-28	kg
Monocrotophos	Emissions to soil	1.2E-11	kg
Monoethanolamine	Emissions to air	2.1E-08	kg
Monoethanolamine	Emission to water	2.5E-12	kg
Monophenyltin	Emission to water	2.9E-31	kg
MSMA	Emissions to soil	2.6E-13	kg
Myclobutanil	Emissions to soil	2.8E-25	kg
Naphtalene	Emissions to air	6.2E-13	kg
Naphtalene	Emission to water	2.5E-13	kg
Naphthalene	Emissions to air	3.7E-17	kg
Napropamide	Emissions to soil	1.4E-12	kg
Nickel	Emissions to air	1.3E-05	kg
Nickel	Emissions to soil	7.5E-09	kg
Nickel, ion	Emission to water	4.0E-05	kg
Nicosulfuron	Emissions to soil	8.4E-16	kg
Niobium-95	Emissions to air	1.2E-05	kBq
Niobium-95	Emission to water	1.8E-08	kBq
Nitrate	Emissions to air	3.1E-08	kg
Nitrate	Emissions to soil	2.9E-09	kg
Nitrate	Emission to water	1.3E-03	kg
Nitric oxide	Emissions to air	2.1E-11	kg
Nitrite	Emission to water	3.8E-08	kg

Nitrobenzene	Emissions to air	1.4E-11	kg
Nitrobenzene	Emission to water	5.5E-11	kg
Nitrogen	Emissions to air	3.4E-05	kg
Nitrogen	Emissions to soil	6.4E-08	kg
Nitrogen	Emission to water	6.6E-07	kg
Nitrogen dioxide, ES	Emissions to air	2.7E-23	kg
Nitrogen fluoride	Emissions to air	3.1E-18	kg
Nitrogen oxides	Emissions to air	7.8E-04	kg
Nitrogen, organic bound	Emission to water	3.1E-05	kg
NMVOC, non-methane volatile organic compounds, unspecified origin	Emissions to air	1.5E-04	kg
Noble gases, radioactive, unspecified	Emissions to air	9.4E-01	kBq
o-Dichlorobenzene	Emission to water	1.2E-09	kg
o-Nitrotoluene	Emissions to air	8.4E-15	kg
o-Xylene	Emissions to air	7.6E-10	kg
o-Xylene	Emission to water	7.9E-11	kg
Oils, non-fossil	Emissions to soil	9.8E-08	kg
Oils, non-fossil	Emission to water	2.2E-10	kg
Oils, unspecified	Emissions to soil	1.2E-05	kg
Oils, unspecified	Emission to water	1.2E-05	kg
Orbencarb	Emissions to soil	1.6E-09	kg
Organic carbon	Emissions to air	2.7E-10	kg
Organic carbon	Emissions to soil	8.7E-10	kg
Organic carbon	Emission to water	8.7E-10	kg
Oxydemeton-methyl	Emissions to soil	5.3E-15	kg
Oxyfluorfen	Emissions to soil	-1.7E-25	kg
Ozone	Emissions to air	2.9E-07	kg
PAH, polycyclic aromatic hydrocarbons	Emissions to air	5.9E-08	kg
PAH, polycyclic aromatic hydrocarbons	Emissions to soil	7.6E-12	kg
PAH, polycyclic aromatic hydrocarbons	Emission to water	1.4E-09	kg
Paraquat	Emissions to air	2.4E-13	kg
Paraquat	Emissions to soil	4.7E-13	kg
Parathion	Emissions to soil	4.2E-13	kg
Particulates, < 2.5 um	Emissions to air	3.4E-04	kg
Particulates, > 10 um	Emissions to air	1.2E-04	kg
Particulates, > 2.5 um, and < 10um	Emissions to air	2.5E-04	kg

Pendimethalin	Emissions to air	2.5E-12	kg
Pendimethalin	Emissions to soil	2.6E-12	kg
Pendimethalin	Emission to water	1.7E-19	kg
Pentane	Emissions to air	6.3E-07	kg
Permethrin	Emissions to air	3.8E-14	kg
Permethrin	Emissions to soil	2.0E-15	kg
Pesticides, unspecified	Emissions to soil	1.3E-11	kg
Phenanthrene	Emissions to air	5.5E-13	kg
Phenanthrene	Emission to water	2.8E-12	kg
Phenmedipham	Emissions to soil	1.4E-15	kg
Phenol	Emissions to air	7.3E-09	kg
Phenol	Emission to water	2.4E-08	kg
Phenol, 2,4-dichloro	Emissions to air	7.6E-14	kg
Phenol, pentachloro-	Emissions to air	1.3E-09	kg
Phenol, pentachloro-	Emissions to soil	1.3E-13	kg
Phosgene	Emissions to air	2.5E-12	kg
Phosphate	Emission to water	4.7E-03	kg
Phosphine	Emissions to air	2.9E-14	kg
Phosphoric acid	Emissions to air	5.6E-18	kg
Phosphorus	Emissions to air	1.5E-08	kg
Phosphorus	Emissions to soil	1.0E-07	kg
Phosphorus	Emission to water	2.7E-08	kg
Phosphorus pentachloride	Emission to water	1.2E-35	kg
Phosphorus trichloride	Emissions to air	8.3E-14	kg
Picloram	Emissions to soil	1.2E-16	kg
Picoxystrobin	Emissions to soil	2.3E-15	kg
Piperonyl butoxide	Emissions to soil	8.1E-25	kg
Pirimicarb	Emissions to soil	1.5E-13	kg
Pirimiphos methyl	Emissions to soil	3.5E-24	kg
Platinum	Emissions to air	1.7E-09	kg
Plutonium-238	Emissions to air	1.3E-14	kBq
Plutonium-alpha	Emissions to air	3.1E-14	kBq
Polonium-210	Emissions to air	4.0E-05	kBq
Polonium-210	Emission to water	2.0E-05	kBq
Polychlorinated biphenyls	Emissions to air	3.0E-11	kg

Polychlorinated biphenyls	Emission to water	4.9E-14	kg
Potassium	Emissions to air	6.4E-07	kg
Potassium	Emissions to soil	5.2E-07	kg
Potassium	Emission to water	1.5E-09	kg
Potassium, ion	Emission to water	1.3E-02	kg
Potassium-40	Emissions to air	7.4E-06	kBq
Potassium-40	Emission to water	4.2E-06	kBq
Primisulfuron	Emissions to soil	3.8E-16	kg
Prochloraz	Emissions to soil	1.7E-14	kg
Procymidone	Emissions to soil	1.9E-13	kg
Profenofos	Emissions to soil	4.0E-13	kg
Prohexadione-calcium	Emissions to soil	7.4E-17	kg
Prometryn	Emissions to soil	2.2E-13	kg
Pronamide	Emissions to soil	2.1E-24	kg
Propanal	Emissions to air	1.6E-10	kg
Propanal	Emission to water	8.3E-13	kg
Propane	Emissions to air	3.5E-07	kg
Propane	Emissions to air	3.4E-07	kg
Propanil	Emissions to soil	3.6E-12	kg
Propanol	Emissions to air	1.9E-12	kg
Propanol	Emission to water	8.0E-13	kg
Propene	Emissions to air	1.3E-07	kg
Propene	Emission to water	2.8E-07	kg
Propiconazole	Emissions to air	4.4E-14	kg
Propiconazole	Emissions to soil	4.8E-14	kg
Propiconazole	Emission to water	9.2E-21	kg
Propionic acid	Emissions to air	3.5E-09	kg
Propionic acid	Emission to water	3.4E-12	kg
Propoxycarbazone-sodium (prop)	Emissions to soil	4.1E-16	kg
Propylamine	Emissions to air	1.4E-13	kg
Propylamine	Emission to water	3.4E-13	kg
Propylene oxide	Emissions to air	2.4E-10	kg
Propylene oxide	Emission to water	5.0E-10	kg
Prosulfuron	Emissions to soil	1.8E-16	kg
Protactinium-234	Emissions to air	3.2E-07	kBq

Protactinium-234	Emission to water	8.3E-07	kBq
Prothioconazol	Emissions to air	3.5E-22	kg
Prothioconazol	Emissions to soil	5.8E-13	kg
Prothioconazol	Emission to water	3.6E-23	kg
Pyraclostrobin	Emissions to air	1.0E-13	kg
Pyraclostrobin	Emission to water	3.8E-20	kg
Pyraclostrobin (prop)	Emissions to soil	1.4E-14	kg
Pyrene	Emissions to air	2.9E-14	kg
Pyrene	Emission to water	2.5E-12	kg
Pyrethrin	Emissions to soil	9.0E-25	kg
Pyrithiobac sodium salt	Emissions to soil	1.4E-14	kg
Quinclorac	Emissions to soil	6.0E-14	kg
Quinoxyfen	Emissions to soil	3.6E-15	kg
Quizalofop ethyl ester	Emissions to soil	2.5E-14	kg
Quizalofop-ethyl	Emissions to air	1.4E-14	kg
Quizalofop-P	Emissions to soil	2.8E-14	kg
Quizalofop-p-ethyl	Emissions to soil	-1.5E-28	kg
Radioactive species, alpha emitters	Emission to water	1.9E-07	kBq
Radioactive species, Nuclides, unspecified	Emission to water	9.6E-05	kBq
Radioactive species, other beta emitters	Emissions to air	3.5E-06	kBq
Radium-224	Emission to water	9.0E-06	kBq
Radium-226	Emissions to air	7.3E-06	kBq
Radium-226	Emission to water	3.1E-04	kBq
Radium-228	Emissions to air	9.3E-06	kBq
Radium-228	Emission to water	2.5E-05	kBq
Radon-220	Emissions to air	1.3E-04	kBq
Radon-222	Emissions to air	3.8E+00	kBq
Rimsulfuron	Emissions to soil	3.8E-16	kg
Rotenone	Emissions to soil	5.0E-25	kg
Rubidium	Emission to water	1.8E-09	kg
Ruthenium-103	Emissions to air	6.8E-13	kBq
Ruthenium-103	Emission to water	1.3E-09	kBq
Scandium	Emissions to air	5.0E-10	kg
Scandium	Emissions to soil	2.5E-10	kg
Scandium	Emission to water	7.8E-06	kg

Selenium	Emissions to air	6.5E-07	kg
Selenium	Emissions to soil	2.8E-10	kg
Selenium	Emission to water	1.6E-05	kg
Sethoxydim	Emissions to air	3.0E-14	kg
Sethoxydim	Emissions to soil	1.3E-13	kg
Silicon	Emissions to air	2.0E-06	kg
Silicon	Emissions to soil	5.1E-06	kg
Silicon	Emission to water	5.7E-03	kg
Silicon dioxide	Emission to water	1.8E-23	kg
Silicon tetrachloride	Emissions to air	2.2E-24	kg
Silicon tetrafluoride	Emissions to air	1.9E-10	kg
Silthiofam	Emissions to soil	5.5E-15	kg
Silver	Emissions to air	4.9E-11	kg
Silver	Emissions to soil	7.8E-11	kg
Silver, ion	Emission to water	1.2E-06	kg
Silver-110	Emissions to air	1.4E-11	kBq
Silver-110	Emission to water	2.4E-06	kBq
Simazine	Emissions to soil	7.7E-15	kg
Sodium	Emissions to air	1.6E-07	kg
Sodium	Emissions to soil	6.3E-07	kg
Sodium	Emission to water	1.2E-09	kg
Sodium chlorate	Emissions to air	1.5E-09	kg
Sodium chlorate	Emission to water	5.9E-14	kg
Sodium dichromate	Emissions to air	2.8E-12	kg
Sodium formate	Emissions to air	8.8E-11	kg
Sodium formate	Emission to water	2.1E-10	kg
Sodium hydroxide	Emissions to air	7.8E-11	kg
Sodium tetrahydridoborate	Emissions to air	2.0E-15	kg
Sodium, ion	Emission to water	4.7E-03	kg
Sodium-24	Emission to water	4.6E-08	kBq
Solids, inorganic	Emission to water	3.0E-05	kg
Spinosad	Emissions to soil	2.8E-23	kg
Spiroxamine	Emissions to soil	2.7E-13	kg
Strontium	Emissions to air	3.9E-08	kg
Strontium	Emissions to soil	1.1E-08	kg

Strontium	Emission to water	3.5E-04	kg
Strontium-89	Emission to water	6.1E-08	kBq
Strontium-90	Emission to water	7.5E-05	kBq
Styrene	Emissions to air	6.6E-08	kg
Sulfate	Emissions to air	1.7E-06	kg
Sulfate	Emissions to soil	4.9E-09	kg
Sulfate	Emission to water	1.4E-01	kg
Sulfate, ion	Emission to water	5.2E-11	kg
Sulfentrazone	Emissions to air	2.9E-13	kg
Sulfentrazone	Emissions to soil	3.6E-13	kg
Sulfide	Emission to water	2.3E-09	kg
Sulfite	Emission to water	5.3E-08	kg
Sulfosate	Emissions to soil	1.4E-12	kg
Sulfosulfuron	Emissions to soil	1.5E-15	kg
Sulfur	Emissions to soil	7.0E-07	kg
Sulfur	Emission to water	7.6E-08	kg
Sulfur dioxide	Emissions to air	5.2E-03	kg
Sulfur hexafluoride	Emissions to air	7.4E-09	kg
Sulfur oxides	Emissions to air	1.3E-09	kg
Sulfur trioxide	Emissions to air	4.4E-11	kg
Sulfuric acid	Emissions to air	3.1E-08	kg
Sulfuric acid	Emissions to soil	3.3E-14	kg
Suspended solids, unspecified	Emission to water	6.5E-06	kg
t-Butyl methyl ether	Emissions to air	2.9E-10	kg
t-Butyl methyl ether	Emission to water	1.5E-11	kg
t-Butylamine	Emissions to air	1.3E-13	kg
t-Butylamine	Emission to water	3.0E-13	kg
tau-Fluvalinate	Emissions to soil	-1.4E-29	kg
Tebuconazole	Emissions to air	9.3E-22	kg
Tebuconazole	Emissions to soil	9.7E-13	kg
Tebuconazole	Emission to water	2.9E-22	kg
Tebupirimphos	Emissions to soil	3.2E-15	kg
Tebutam	Emissions to soil	4.4E-14	kg
Technetium-99m	Emission to water	7.8E-08	kBq
Teflubenzuron	Emissions to soil	1.9E-11	kg

Tefluthrin	Emissions to air	9.1E-20	kg
Tefluthrin	Emissions to soil	2.7E-15	kg
Tefluthrin	Emission to water	4.5E-25	kg
Tellurium-123m	Emission to water	8.9E-09	kBq
Tellurium-132	Emission to water	3.7E-10	kBq
Terbacil	Emissions to soil	7.2E-23	kg
Terbufos	Emissions to soil	9.4E-15	kg
Terpenes	Emissions to air	6.7E-11	kg
Tetramethyl ammonium hydroxide	Emissions to air	7.4E-14	kg
Thallium	Emissions to air	2.1E-10	kg
Thallium	Emissions to soil	2.3E-11	kg
Thallium	Emission to water	2.0E-06	kg
Thiamethoxam	Emissions to soil	2.5E-14	kg
Thidiazuron	Emissions to soil	2.5E-14	kg
Thifensulfuron	Emissions to air	4.1E-15	kg
Thifensulfuron-methyl	Emissions to soil	1.3E-15	kg
Thiobencarb	Emissions to soil	7.7E-13	kg
Thiodicarb	Emissions to air	1.5E-14	kg
Thiodicarb	Emissions to soil	6.3E-16	kg
Thiram	Emissions to soil	1.9E-12	kg
Thorium	Emissions to air	2.3E-10	kg
Thorium-228	Emissions to air	1.5E-06	kBq
Thorium-228	Emission to water	3.6E-05	kBq
Thorium-230	Emissions to air	6.4E-07	kBq
Thorium-230	Emission to water	7.0E-05	kBq
Thorium-232	Emissions to air	1.6E-06	kBq
Thorium-232	Emission to water	5.1E-07	kBq
Thorium-234	Emissions to air	3.2E-07	kBq
Thorium-234	Emission to water	8.3E-07	kBq
Tin	Emissions to air	8.2E-07	kg
Tin	Emissions to soil	4.8E-09	kg
Tin, ion	Emission to water	2.0E-05	kg
Titanium	Emissions to air	5.5E-08	kg
Titanium	Emissions to soil	9.3E-08	kg
Titanium, ion	Emission to water	2.5E-05	kg

TOC, Total Organic Carbon	Emission to water	3.0E-04	kg
Toluene	Emissions to air	4.7E-07	kg
Toluene	Emission to water	2.9E-08	kg
Toluene, 2-chloro	Emissions to air	2.9E-12	kg
Toluene, 2-chloro	Emission to water	6.9E-12	kg
Tralkoxydim	Emissions to soil	4.3E-16	kg
Tri-allate	Emissions to soil	3.3E-15	kg
Triadimenol	Emissions to soil	3.6E-15	kg
Triasulfuron	Emissions to soil	9.8E-16	kg
Tribenuron	Emissions to soil	2.8E-16	kg
Tribenuron-methyl	Emissions to soil	1.2E-15	kg
Tribufos	Emissions to soil	2.4E-13	kg
Tributyltin compounds	Emission to water	9.5E-10	kg
Trichloroethylene	Emissions to air	2.3E-11	kg
Triclopyr	Emissions to soil	3.1E-11	kg
Triethylene glycol	Emission to water	8.6E-10	kg
Trifloxystrobin	Emissions to air	2.6E-15	kg
Trifloxystrobin	Emissions to soil	4.8E-15	kg
Trifloxystrobin	Emission to water	1.1E-23	kg
Trifluralin	Emissions to air	4.1E-12	kg
Trifluralin	Emissions to soil	6.8E-12	kg
Trimethylamine	Emissions to air	4.7E-15	kg
Trimethylamine	Emission to water	1.1E-14	kg
Trinexapac-ethyl	Emissions to soil	2.9E-14	kg
Trioctyltin	Emission to water	6.8E-29	kg
Triphenyltin	Emission to water	2.9E-29	kg
Tungsten	Emissions to air	4.0E-11	kg
Tungsten	Emission to water	2.9E-05	kg
Uranium	Emissions to air	3.0E-10	kg
Uranium alpha	Emissions to air	2.0E-06	kBq
Uranium alpha	Emission to water	3.2E-05	kBq
Uranium-234	Emissions to air	1.2E-06	kBq
Uranium-234	Emission to water	9.6E-07	kBq
Uranium-235	Emissions to air	1.7E-08	kBq
Uranium-235	Emission to water	1.1E-06	kBq

Uranium-238	Emissions to air	5.6E-06	kBq
Uranium-238	Emission to water	9.1E-06	kBq
Urea	Emission to water	9.6E-13	kg
Vanadium	Emissions to air	1.3E-07	kg
Vanadium	Emissions to soil	3.2E-09	kg
Vanadium, ion	Emission to water	1.0E-05	kg
Vinclozolin	Emissions to soil	6.5E-14	kg
VOC, volatile organic compounds	Emissions to air	1.4E-22	kg
VOC, volatile organic compounds, unspecified origin	Emissions to air	3.2E-16	kg
VOC, volatile organic compounds, unspecified origin	Emission to water	6.5E-08	kg
Water	Emissions to air	1.7E-03	m3
Water	Emission to water	1.4E+00	m3
Xenon-131m	Emissions to air	5.7E-05	kBq
Xenon-133	Emissions to air	3.4E-03	kBq
Xenon-133m	Emissions to air	2.2E-06	kBq
Xenon-135	Emissions to air	1.2E-03	kBq
Xenon-135m	Emissions to air	5.2E-04	kBq
Xenon-137	Emissions to air	1.6E-05	kBq
Xenon-138	Emissions to air	1.2E-04	kBq
Xylene	Emissions to air	4.1E-07	kg
Xylene	Emission to water	2.1E-08	kg
Zeta-cypermethrin	Emissions to air	1.7E-14	kg
Zeta-cypermethrin	Emissions to soil	7.4E-16	kg
Zinc	Emissions to air	6.4E-06	kg
Zinc	Emissions to soil	1.0E-07	kg
Zinc, ion	Emission to water	1.1E-03	kg
Zinc-65	Emissions to air	1.3E-10	kBq
Zinc-65	Emission to water	1.1E-06	kBq
Zirconium	Emissions to air	1.9E-13	kg
Zirconium-95	Emissions to air	2.6E-10	kBq
Zirconium-95	Emission to water	5.0E-06	kBq

Appendix B: Environmental Impacts of all THEVA Tape Components

Environmental impacts of materials used within the inclined substrate deposition process based on the Environmental Footprint 3.0 impact assessment method and the cumulative energy demand.

Impact category	Unit	Electrolyte supplement 1 kg	Heat (borehole heat pump) 1 kg	Polypropylene 1 kg	Transport 1 t*km	Water 1 kg
Acidification terrestrial and freshwater	mol H+ eq.	0.02	2.60E-04	0.01	0.01	2.52E-06
Cancer human health effects	CTUh	2.69E-07	4.63E-10	1.35E-08	2.82E-08	3.85E-11
Climate change	kg CO_2eq.	1.37	0.04	2.20	1.86	3.60E-04
Ecotoxicity freshwater	CTUe	4.87	0.01	0.92	0.84	5.30E-04
Eutrophication freshwater	kg Peq.	1.53E-03	3.58E-05	7.49E-05	4.00E-04	2.77E-07
Eutrophication marine	kg Neq.	2.27E-03	3.71E-05	1.52E-03	3.14E-03	4.08E-07
Eutrophication terrestrial	mol Neq.	0.02	5.80E-04	0.02	0.04	5.58E-06
Ionising radiation	kBq U-235eq.	0.14	0.02	0.01	0.18	1.30E-04
Land use	Pt.	74.29	0.03	1.10	13.76	4.60E-04
Non-cancer human health effects	CTUh	8.76E-07	4.39E-09	2.12E-08	1.70E-07	1.64E-10
Ozone depletion	kg CFC-11eq.	2.11E-07	6.26E-09	1.78E-08	3.94E-07	3.07E-11
Photochemical ozone formation	kg NMVOCeq.	0.01	8.65E-05	0.01	0.01	1.22E-06
Resource use, energy carriers	MJ	24.35	0.68	70.90	28.19	0.01
Resource use, mineral and metals	kg Sbeq.	3.14E-05	7.68E-08	2.40E-07	7.13E-06	1.04E-09
Respiratory inorganics	Disease incidences.	1.53E-07	1.09E-09	7.87E-08	1.87E-07	2.01E-11
Water scarcity	m^3 deprived	214.02	16.42	6.88	127.94	0.17
Cumulative energy demand	kWh	8.00	0.44	21.22	8.49	2.0E-03

Appendix C: Environmental Impacts of all Oxolutia Tape Components

Environmental impacts of the main materials used within the inkjet printing process based on the Environmental Footprint 3.0 impact assessment method and the cumulative energy demand.

Impact category	Unit	1-butanol 1 kg	Diethanolamine 1 kg	Fleece polyethylene 1 kg	Isopropanol 1 kg	Propionic acid 1 kg	Triethanolamine 1 kg
Acidification terrestrial and freshwater	mol H+ eq.	0.02	0.01	0.01	0.01	0.01	0.01
Cancer human health effects	CTUh	2.97	2.90	3.13	2.09	2.04	2.95
Climate change	kg CO_2eq.	45.19	33.17	13.63	14.36	29.92	31.41
Ecotoxicity freshwater	CTUe	8.3E-04	7.9E-04	2.9E-04	3.1E-04	7.2E-04	8.2E-04
Eutrophication freshwater	kg Peq.	2.4E-03	0.01	2.4E-03	1.2E-03	1.7E-03	0.01
Eutrophication marine	kg Neq.	0.03	0.03	0.03	0.01	0.02	0.03
Eutrophication terrestrial	mol Neq.	7.5E-10	9.3E-09	7.3E-10	4.2E-10	6.1E-10	1.0E-08
Ionising radiation	kBq U-235eq.	3.3E-08	2.6E-08	2.7E-08	1.1E-08	1.9E-08	2.6E-08
Land use	Pt.	0.24	0.22	0.07	0.10	0.26	0.23
Non-cancer human health effects	CTUh	1.40	0.41	1.71	0.50	1.60	0.28
Ozone depletion	kg CFC-11eq.	3.4E-07	1.5E-07	8.6E-08	9.9E-08	3.0E-07	1.3E-07
Photochemical ozone formation	kg NMVOCeq.	1.4E-07	1.1E-07	1.4E-07	7.6E-08	7.6E-08	1.0E-07
Resource use, energy carriers	MJ	0.01	0.01	0.01	0.01	0.01	0.01
Resource use, mineral and metals	kg Sbeq.	77.01	68.05	89.24	59.58	55.29	70.42
Respiratory inorganics	Disease incidences.	6.9E-06	9.7E-06	5.2E-06	5.2E-06	6.7E-06	9.8E-06
Water scarcity	m^3 deprived	344.78	381.25	106.73	101.30	360.08	396.28
Cumulative energy demand	kWh	23.50	20.80	27.11	18.00	16.86	21.52

Appendix D: Life Cycle Inventories of Components of an Open Cooling System

Life cycle inventory of the production of a circulation pump.

Material	Input Amount	Unit
Alkyd paint, white	1.3	kg
Aluminium, cast alloy	3.0	kg
Aluminium, wrought alloy	0.9	kg
Bronze	4.7	kg
Cast iron	76.7	kg
Casting, bronze	4.7	kg
Copper	10.8	kg
Electronics	6.6	kg
Epoxy resin	1.7	kg
Hot water tank factory	2.0E-7	#
Metal working	141.6	kg
Polyvinylchloride	3.4	kg
Printed wiring board	0.9	kg
Scrap steel	-239.2	kg
Silicon carbide	0.8	kg
Steel, chromium steel	239.2	kg
Transport, freight	19.6	t*km
Wire drawing, copper	10.8	kg
Material	**Output Amount**	**Unit**
Circulation pump	1	#

Life cycle inventory of the production of a liquid nitrogen storage tank.

Material	Input Amount	Unit
Alkyd paint, white	68.8	kg
Perlite	2776.4	kg
Scrap steel	-2.3E4	kg
Sheet rolling, chrome	1.0E4	kg
Sheet rolling, steel	9674.0	kg
Steel, chromium steel	1.0E4	kg
Steel, low-alloyed	9674.0	kg
Transport, freight	1829.6	t*km
Welding, arc, steel	48.42	m
Material	**Output Amount**	**Unit**
Liquid nitrogen storage tank	1	#

Life cycle inventory of the production of piping.

Material	Input Amount	Unit
Metal working	129.3	kg
Polyurethane	20.0	kg
Scrap steel	-129.3	kg
Steel, chromium	129.3	kg
Transport, freight	32.3	t*km
Material	**Output Amount**	**Unit**
Piping	1	#

Life cycle inventory of the production of a subcooler.

Material	Input Amount	Unit
Copper	250.0	kg
Electricity, medium voltage	366.8	MJ
Extrusion	250.0	kg
Scrap steel	-250.0	kg
Sheet rolling, chromium	250.0	kg
Steel, chromium	250.0	kg
Transport, freight	121	t*km
Ventilation	2.6E-7	#
Welding, arc, steel	6	m
Wire drawing, copper	250.0	kg
Material	**Output Amount**	**Unit**
Subcooler	1	#

Life cycle inventory of the production of a vacuum pump.

Material	Input Amount	Unit
Alkyd paint, white	0.5	kg
Aluminium, cast alloy	9.2	kg
Aluminium, wrought alloy	13.8	kg
Cast iron	68.5	kg
Copper	29.7	kg
Epoxy resin	1.5	kg
Metal working	90.1	kg
Polyvinylchloride	0.5	kg
Printed wiring board	1.0	kg
Scrap steel	-160.2	kg
Sheet rolling, aluminium	20.3	kg
Sheet rolling, chromium	70.1	kg
Steel, chromium	160.2	kg
Transport, freight	116.9	t*km
Wire drawing, copper	29.7	kg
Material	**Output Amount**	**Unit**
Vacuum pump	1	#

Appendix E: Impact Factors of Selected Metals in the Impact Category Resource Use of Minerals and Metal

Impact factor of all metals that are considered in Figure 4.20 for the impact category resource use of minerals and metals.

Metal	Impact factor	Unit
Cadmium	0.157	kg Sb equivalents / kg
Chromium	4.43E-4	kg Sb equivalents / kg
Copper	0.00137	kg Sb equivalents / kg
Gold	52.0	kg Sb equivalents / kg
Lead	0.00634	kg Sb equivalents / kg
Molybdenum	0.0178	kg Sb equivalents / kg
Nickel	6.53E-5	kg Sb equivalents / kg
Silver	1.18	kg Sb equivalents / kg
Antimony (reference flow)	1.0	kg Sb equivalents / kg

Appendix F: Life Cycle Inventories of Components of a Closed Cooling System

Life cycle inventory of the production of a cryocooler.

Material	Input Amount	Unit
Alkyd paint, white	1.6	kg
Aluminium alloy	55.4	kg
Cast iron	115.7	kg
Casting, aluminium	26.4	kg
Casting, steel	382.7	kg
Copper	172.6	kg
Electronics	23.0	kg
Epoxy resin	4.5	kg
Helium	1.3	kg
Metal working	701.2	kg
Polyvinylchloride	1.6	kg
Scrap steel	-659.5	kg
Steel, chromium	659.5	kg
Steel, low-alloyed	115.0	kg
Transport, freight	153.0	t*km
Wire drawing, copper	172.6	kg
Material	**Output Amount**	**Unit**
Cryocooler	1	#

Life cycle inventory of the production of a phase separator.

Material	Input Amount	Unit
Scrap steel	-40.0	kg
Steel, chromium	40.0	kg
Transport, freight	3.2	t*km
Welding, arc, steel	1.4	m
Material	**Output Amount**	**Unit**
Phase separator	1	#

Life cycle inventory of the production of a vacuum insulated cold box.

Material	Input Amount	Unit
Scrap steel	-277.0	kg
Sheet rolling, steel	277.0	kg
Steel, low-alloyed	277.0	kg
Transport, freight	36.8	t*km
Welding, arc, steel	1.4	m
Material	**Output Amount**	**Unit**
Vacuum insulated cold box	1	#

Life cycle inventory of the production of valves.

Material	Input Amount	Unit
Bronze	3.0	kg
Casting, bronze	3.0	kg
Casting, steel	257.9	kg
Scrap steel	-257.9	kg
Steel, low-alloyed	257.9	kg
Transport, freight	13.8	t*km
Material	**Output Amount**	**Unit**
Valves	1	#

Karlsruher Schriftenreihe zur Supraleitung

Karlsruher Institut für Technologie (KIT) | ISSN 1869-1765

Band 001 **Christian Schacherer**
Theoretische und experimentelle Untersuchungen zur Entwicklung supraleitender resistiver Strombegrenzer.
ISBN 978-3-86644-412-6

Band 002 **Alexander Winkler**
Transient behaviour of ITER poloidal field coils.
ISBN 978-3-86644-595-6

Band 003 **André Berger**
Entwicklung supraleitender, strombegrenzender Transformatoren.
ISBN 978-3-86644-637-3

Band 004 **Christoph Kaiser**
High quality Nb/Al-AlOx/Nb Josephson junctions. Technological development and macroscopic quantum experiments.
ISBN 978-3-86644-651-9

Band 005 **Gerd Hammer**
Untersuchung der Eigenschaften von planaren Mikrowellen-resonatoren für Kinetic-Inductance Detektoren bei 4,2 K.
ISBN 978-3-86644-715-8

Band 006 **Olaf Mäder**
Simulationen und Experimente zum Stabilitätsverhalten von HTSL-Bandleitern.
ISBN 978-3-86644-868-1

Band 007 **Christian Barth**
High Temperature Superconductor Cable Concepts for Fusion Magnets.
ISBN 978-3-7315-0065-0

Band 008 **Axel Stockhausen**
Optimization of Hot-Electron Bolometers for THz Radiation.
ISBN 978-3-7315-0066-7

Band 009 **Petra Thoma**
Ultra-fast $YBa_2Cu_3O_{7-x}$ direct detectors for the THz frequency range.
ISBN 978-3-7315-0070-4

Band 010 **Dagmar Henrich**
Influence of Material and Geometry on the Performance of Superconducting Nanowire Single-Photon Detectors.
ISBN 978-3-7315-0092-6

Band 011 **Alexander Scheuring**
Ultrabreitbandige Strahlungseinkopplung in THz-Detektoren.
ISBN 978-3-7315-0102-2

Band 012 **Markus Rösch**
Development of lumped element kinetic inductance detectors for mm-wave astronomy at the IRAM 30 m telescope.
ISBN 978-3-7315-0110-7

Band 013 **Johannes Maximilian Meckbach**
Superconducting Multilayer Technology for Josephson Devices.
ISBN 978-3-7315-0122-0

Band 014 **Enrico Rizzo**
Simulations for the optimization of High Temperature Superconductor current leads for nuclear fusion applications.
ISBN 978-3-7315-0132-9

Band 015 **Philipp Krüger**
Optimisation of hysteretic losses in high-temperature superconducting wires.
ISBN 978-3-7315-0185-5

Band 016 **Matthias Hofherr**
Real-time imaging systems for superconducting nanowire single-photon detector arrays.
ISBN 978-3-7315-0229-6

Band 017 **Oliver Näckel**
Development of an Air Coil Superconducting Fault Current Limiter.
ISBN 978-3-7315-0526-6

Band 018 **Christoph M. Bayer**
Characterization of High Temperature Superconductor Cables for Magnet Toroidal Field Coils of the DEMO Fusion Power Plant.
ISBN 978-3-7315-0605-8

Band 019 **Shengnan Zou**
Magnetization of High Temperature Superconducting Trapped-Field Magnets.
ISBN 978-3-7315-0715-4

Band 020 **Ilya Charaev**
Improving the Spectral Bandwidth of Superconducting Nanowire Single-Photon Detectors (SNSPDs).
ISBN 978-3-7315-0745-1

Band 021 **Juliane Raasch**
Electrical-field sensitive $YBa_2Cu_3O_{7-x}$ detectors for real-time monitoring of picosecond THz pulses.
ISBN 978-3-7315-0786-4

Band 022 **Yingzhen Liu**
Design of a superconducting DC wind generator.
ISBN 978-3-7315-0796-3

Band 023 **Sebastian Hellmann**
Research and Technology Development on Superconducting Current Limiting Transformers.
ISBN 978-3-7315-0804-5

Band 024 **Simon J. Otten**
Characterisation of REBCO Roebel cables.
ISBN 978-3-7315-0904-2

Band 025 **Julia Brandel**
Supraleitende Einzelphotonenzähler: Optimierung der Zeitauflösung und Anwendungsbeispiele aus der Spektroskopie.
ISBN 978-3-7315-0917-2

Band 026 **Dustin Kottonau, Eugen Shabagin, Wescley T. B. de Sousa, Jörn Geisbüsch, Mathias Noe, Hanno Stagge, Simon Fechner, Hannes Woiton, Thomas Küsters**
Bewertung des Einsatzes supraleitender 380-kV-Kabel.
ISBN 978-3-7315-0927-1

Band 027 **Steffen Dörner**
Multifrequenzausleseverfahren von supraleitenden Einzelphotonen-Detektoren.
ISBN 978-3-7315-0961-5

Band 028 **Michael Merker**
Superconducting integrated THz receiver.
ISBN 978-3-7315-0970-7

Band 029 **Wolfgang-Gustav Ekkehart Schmidt**
Superconducting Nanowire Single-Photon Detectors for Quantum Photonic Integrated Circuits on GaAs.
ISBN 978-3-7315-0980-6

Band 030 **Dustin Kottonau, Eugen Shabagin, Wescley de Sousa, Jörn Geisbüsch, Mathias Noe, Hanno Stagge, Simon Fechner, Hannes Woiton, Thomas Küsters**
Evaluation of the Use of Superconducting 380 kV Cable.
ISBN 978-3-7315-1026-0

Band 031 **Alan Preuß**
Development of high-temperature superconductor cables for high direct current applications.
ISBN 978-3-7315-1041-3

Die Bände sind unter www.ksp.kit.edu als PDF frei verfügbar oder als Druckausgabe bestellbar.

Band 032 **Roland Gyuráki**
Fluorescent thermal imaging method for investigating transient effects in high-temperature superconductor tapes and coils.
ISBN 978-3-7315-1064-2

Band 033 **Aurélien Godfrin**
AC Losses in High-Temperature Superconductor Tapes and Cables for Power Applications.
ISBN 978-3-7315-1096-3

Band 034 **Alexander Schmid**
Integrierte planare Antennen für supraleitende Detektorsysteme zur THz-Strahldiagnostik.
ISBN 978-3-7315-1145-8

Band 035 **Alexander Buchholz**
Prospective Life Cycle Assessment of High-Temperature Superconductors for Future Grid Applications.
ISBN 978-3-7315-1194-6